Advanced Series in Agricultural Sciences

Advanced Series in Agricultural Sciences

Volumes already published in the series

Volume 10
Saline and Sodic Soils (1982)
E. Bresler, B. L. McNeal, D. L. Carter

Volume 11
A Theory of Feeding and Growth of Animals (1982)
J. R. Parks

Volume 12
Fertilization of Dryland and Irrigated Soils (1982)
J. Hagin, B. Tucker

Volume 13
Agricultural Soil Mechanics (1983)
A. J. Koolen, H. Kuipers

Volume 14
Energy and Agriculture (1984)
G. Stanhill (Ed.)

Volume 15
The Rhizosphere (1986)
E. A. Curl, B. Truelove

Volume 16
Population Genetics: Basic Principles (1987)
D. P. Doolittle

Volume 17
Irrigation with Treated Sewage Effluent (1991)
A. Feigin, I. Ravina, J. Shalhevet

Volume 18
Advances in Statistical Methods for Genetic Improvement of Livestock (1990)
D. Gianola, K. Hammond (Eds.)

Volume 19
Electromagnetic Radiations in Food Science (1992)
I. Rosenthal

Volume 20
Water Flow and Solute Transport in Soils (1993)
D. Russo, G. Dagan (Eds.)

Volume 21
Production of Sugarcane: Theory and Practice (1993)
G. Hunsigi

Volume 22
Management of Water Use in Agriculture (1994)
K. K. Tanji, B. Yaron (Eds.)

Volume 23
Woody Plants in Agro-Ecosystems of Semi-Arid Regions (1995)
H. Breman, J.-J. Kessler

Volume 24
Soilless Culture Management (1995)
M. Schwarz

Meier Schwarz

Soilless Culture Management

With 30 Figures and 21 Tables

Springer-Verlag
Berlin Heidelberg New York
London Paris Tokyo
Hong Kong Barcelona
Budapest

Prof. Meier Schwarz, Ph.D.
Bone Hachoma
Old City, Jerusalem 97500
Israel
and
Jerusalem College of Technology
P.O.B. 16031
Jerusalem 91160
Israel

ISBN 3-540-58159-6 Springer-Verlag Berlin Heidelberg New York
ISBN 0-387-58159-6 Springer-Verlag New York Berlin Heidelberg

Library of Congress Cataloging-in-Publication Data. Schwarz, M. (Meier), 1926– . Soilless culture management / M. Schwarz. p. cm. – (Advanced series in agricultural sciences; v. 24) Includes bibliographical references (p.) and index. ISBN 3-540-58159-6 (Springer-Verlag Berlin Heidelberg New York). – ISBN 0-387-58159-6 (Springer-Verlag New York Berlin Heidelberg). 1. Hydroponics. I. Title. II. Series: Advanced series in agricultural sciences; 24. SB126.5.S33 1995 631.5′85 – dc20 94-38472

Typesetting: Macmillan India Ltd, Bangalore 25

SPIN: 10056964 31/3130/SPS – 5 4 3 2 1 0 – Printed on acid-free paper

Foreword

The method of growing plants without soil, known as hydroponics, has long been a subject of much public interest in Israel. Israel, short on water supply and limited in agricultural soil, was looking for alternatives.

The study of hydroponics in Israel received additional stimulus from a somewhat unrelated source. Owing to observance of the Sabbatical year (which required the land to be left follow every seventh year), certain religious settlements in Israel were faced with the problem of providing fresh vegetables for their members during that seventh year. In the course of seeking solutions to this problem, practical experiments with the gravel culture method began in 1952 at Kibbutz Hafetz-Hayim, of which the author was a member.

Although the method was initially introduced for religious reasons, it soon became apparent that hydroponic culture might be an economically viable means of growing flowers and vegetables under the special conditions prevailing in Israel.

In the Negev, the desert area of Israel, arable land and fresh water are scarce, and readily available water is brackish. A hope arose that, thanks to this method, less water would be required and brackish water could be utilized to enable development of the region. Research was carried out at the Negev Institute for Arid Zone Research in Beersheva, including field trials at desert settlements.

The development of new and more economic methods, especially sand culture and rockwool culture, opens most regions for economic soilless culture. More than 30% of certain vegetables and flowers have been produced recently in Israel by these methods.

The awaited blooming of the deserts will be fundamentally, connected with soilless culture methods. Soilless culture is already the future.

In common with many countries around the world, in arid regions under controlled glasshouse conditions in agricultural regions, and in space programs, salinity problems have contributed to the enormous progress in this field of plant production.

Direct contact to many Agriculture Institutes, to research, as well as to agriculturists and their experience, enable us to present this subject.

My thanks to International Groups such as The International Society of Soilless Culture in the Netherlands, the Hydroponic Society in the USA and in Germany, where the exchange of information stimulated new ideas. Special thanks to the Plant Protection and Production Department of the Food and Agriculture Organization (FAO) of UN in Rome, who opened the soilless culture methods to the Third World.

I thank also my colleagues at the Jerusalem College of Technology, where our recent activities took place.

Preface

In selecting material to be included in this book, emphasis has been placed on recent developments in rapidly expanding areas of applied research, and, new methods and ideas in modern agriculture.

The book is intended to serve as a guide for students in the agricultural sciences, for agriculture instructors and for the soilless-culture farmer.

The needs of the commercial soilless farmer fall between those of controlled scientific plant growth and those of ordinary agriculture. Advanced understanding of plant growth, nutrition and environment requirements, and their limitations, will provide the farmer with improved capability for observation of his crops.

With the improved understanding that the farmer will gain about his plants from this book, he will be better able to ask more precise and pointed questions of the consultant.

Jerusalem, Israel
December 1994

M. SCHWARZ

Contents

1 History 1

2 Types of Hydroponics and Nomenclature 3
2.1 Advantages and Disadvantages 4

3 Nutrients 7
3.1 The Elemental Composition of Plants 7
3.2 Macroelements 9
3.3 Microelements 14
3.4 Nonrequired Elements 17
3.5 Deficiency Symptom Considerations 18
3.6 Interaction Between Nutrients 19
3.7 Methods of Expressing Solution Concentration 20
3.8 Composition of the Nutrient Solution 22
3.9 Mixed Fertilizers 27
3.10 pH 28
3.11 Mobile and Immobile Nutrients 28
3.12 Nutrient Concentration and Yield 29
3.13 Preparation of Nutrient Concentrates 29
3.14 Foliar Nutrition 30
3.15 Nutrient Solution Analysis 31

4 Culture Methods 33
4.1 General Outlines 33
4.2 The Supporting Media 34
4.3 Sand Culture – Open System 40
4.4 Rockwool Culture – Open System 45
4.5 Gravel Culture – Closed System 56
4.6 Nutrient Film Technique (NFT) and Water Culture – Closed System 66
4.7 Water Culture – Closed System 77
4.8 Bag Culture – Open Systems 83
4.9 Other Systems 90

5 Water 92
5.1 General 92

5.2 The Process of Osmosis 93
5.3 Transpiration 93
5.4 Irrigation 96
5.5 Saline Water 96
5.6 Wastewater (Sewage) 104

6 Carbon Dioxide 108
6.1 General Observations 109
6.2 CO_2 Sources 113
6.3 Economics of CO_2 Enrichment 116

7 Root Environment 117
7.1 Root Functioning 117
7.2 Relations of Nutrient Absorption 119

8 Plant Growth Effects and Plant Tests 123
8.1 Photosynthetic Efficiency 123
8.2 Harvest Index 123
8.3 Root-Shoot Relationships 125
8.4 Plant Tests: Choice of Tissue 125
8.5 Plant Analysis 127
8.6 Disease Control 129

9 Controlling Systems 131
9.1 Controllers 131
9.2 Temperature 135
9.3 Light 138
9.4 Automated Fertilizer Application 143
9.5 EC Controller (Conductivity) 144
9.6 CO_2 Controllers 145
9.7 DO Controller 145

10 Culture and Cultivation 147
10.1 Plants Adaptable to Hydroponic Culture 147
10.2 Tomato 150
10.3 Cucumber 153
10.4 Lettuce 155
10.5 Flowers 156

11 Establishing a Commercial Soilless Culture Unit 158
11.1 Economic Considerations 159
11.2 Increased Technical Demands on the Management 161

12 Some Future Aspects 162
12.1 Plant Nutrients 162
12.2 Initial Investment 162
12.3 Growth Media 163

12.4 Advanced Sand-Culture Methods 163

References 165

Appendix 171

Subject Index... 193

1 History

The early history of hydroponics (nutriculture, or soilless culture of plants) is closely interwoven with that of plant physiology. In 1600, Jan Van Helmont of Belgium conducted his classical experiment, which, he claimed, proved that plants obtain all their substances from water. Van Helmont planted a 50-lb willow shoot in a tube containing 200 lbs of dried soil, which was covered to keep out dust. After 5 years of regularly adding only rainwater to the earth in the tube, the shoot increased in weight by 160 lbs. The dried soil lost less than 2 ounces in weight, a fact which Van Helmont believed to be insignificant. The plant, of course, did not live on the rainwater alone; we know today that it obtained many elements from the soil, as well as carbon dioxide and oxygen from the air.

In 1699, John Woodward of England grew plants in water to which soil had been added, and he found that the most marked growth occurred in water containing moist soil. In France, about 100 years later, De Saussure and Boussingault showed that plants contain carbon, hydrogen, oxygen, and nitrogen. Sachs and Knop in Germany in 1860 found that, in addition to the above four elements, which are obtained from air, water, or soil, all plants also require phosphorus, sulfur, potassium, calcium, and magnesium. They then grew plants in water solution containing only the salts of these added elements.

Since then, with the refinement of chemicals and of experimental techniques, the need for iron, zinc, copper, boron, manganese, molybdenum, and chlorine has been established. These are called microelements due to the minute quantities required, which, indeed, led to difficulties in the initial experiments, since they were generally present in these amounts as impurities in commercial chemicals.

Once it was known that plants could be grown in solutions of water plus chemicals (nutrients), hydroponic culture was scientifically used to establish the optimal nutrient requirements for various plants. Some attempts at commercial water culture were made in Switzerland and the USA between the two World Wars, but they generally proved uneconomic. The Second World War rendered many barren islands in the Atlantic and Pacific Oceans strategically important, giving impetus to the development of gravel culture methods; some hydroponic gardens supplied fresh vegetables to the military and civilian personnel stationed on those islands (Jones 1982).

Today, hydroponic gardens supply vegetables and flowers under various climate conditions, in Europe under glasshouse conditions, especially in Italy, France, The Netherlands, Germany, England, and the former USSR; in the USA, Japan, Canada, and in Israel, especially in arid zones; under more tropical conditions, in Singapore, India, Kuwait, and the Canary Islands (Schwarz 1968; Steiner 1976).

Thus, what began 100 years ago as an experimental tool in plant physiology has today become a method for growing plants in areas or under economic conditions previously considered highly unfavorable.

Man, striving to produce more and better food plants, has attempted to alter the biosphere. These attempts have generally been directed at changing conditions in the growth medium, and have comprised plowing to expose, aerate, and mix the soil and its constituents, the addition of nutrient elements in the form of organic and inorganic fertilizers, and the provision of irrigation and drainage to create optimal water and air ratios. In cooler climates, man has also attempted to change conditions of temperature, humidity, and even light by growing plants in glass- or hothouses, as a means of industrializing plant growth.

By the use of hydroponics (from hydros = having to do with water + ponos = labor) the biosphere is altered – not by changing some of the conditions in the growth medium or the immediate plant environment – but by completely changing the growth medium and thereby eliminating all dependence on soil. It is thus necessary to find ways of replacing the two major functions of soil – as a support for the plant, and as a source of water and nutrients for plant growth. In all hydroponic systems, the nutrition problem is solved in a similar fashion: nutrient elements are placed in solution in the amounts and proportions required by various plants and the solution is brought into direct contact with the plant roots – hence the common name.

2 Types of Hydroponics and Nomenclature

"Soilless culture" covers all methods and systems of growing plants without soil, such as water culture, sand culture, gravel culture, aeroponics, tube culture, nutriculture, etc. The word hydroponics was originally used in the United States as a synonym for soilless culture. Likewise, the German Hydrokultur and the English hydroculture have been used to describe this method of farming (Steiner 1976; Jones 1983; FAO 1990).

The various soilless culture systems now in use may be grouped according to their type of support medium, the size of the support medium (more or less than 3 mm), and installation (Gericke 1937, 1938; Davtyan 1980).

1. Liquid or Water (Tank) Culture. The plant is supported above the roots on cardboard, plastic, wood or wire; the roots are continuously or intermittently submerged in a nutrient solution or nutrient solution film (Ellis and Swaney 1947).

This system includes tube culture, nutrient film techniques (NFT), and flooding hydroponics. In most of these systems the nutrient solution is reused.

2. Sand Culture. The roots of the plants are grown in a solid substrate consisting of porous or nonporous, noncollapsing particles (sand, perlite, plastics, rockwool, or any other inorganic material) having a diameter of less than 3 mm, primarily utilizing aboveground irrigation.

3. Gravel Culture. The roots of the plant are grown in a solid substrate consisting of porous or nonporous, noncollapsing particles (gravel, basalt, pumice, lava, or any other inorganic material) having a diameter greater than 3 mm.

This culture is intended for the following methods:

a) Tubirrigation method: the nutrient solution, which is stored in an underground reservoir, is pumped into the beds. When the beds are filled to the required level, the pump is stopped, the solution is drained back into the reservoir, and air fills the spaces between the gravel particles.
b) Surface irrigation method: using sand or gravel support media [as in (2) or (3)], a diluted nutrient solution is distributed onto the surface of the growth medium by means of special perforated pipes. The

nutrients are introduced into the main water pipe system with a liquid fertilizer apparatus. After the growth medium has been moistened, the unabsorbed nutrient solution is drained. Air is sucked in during the intervals between irrigation.

4. Aeroponics. The roots of the plants are located in an environment continuously or intermittently saturated with fine drops (a mist of aerosol) of nutrient solution.

Other special growth media are:

Vermiculaponics: the roots of the plants are grown in vermiculite, sometimes mixed with any other inorganic material.
Rockwool culture: the roots of the plants are grown in rockwool or allied inorganic compounds, e.g., glasswool.
Hydroculture: all methods and systems of soilless culture, especially if used for growth of ornamental plants in homes and offices.
Plastoponics: the roots of the plants are grown in expanded plastic.

Growth in organic substrates such as peat, composts, sawdust, and mixtures of organic and nonorganic substrates.

Other terms for soilless culture found in the literature are media culture, soilless media culture, substrate culture, aggregate culture, and chemiculture.

An important classification is often made by differentiating between:

a) Open systems, where the nutrient solution is not reused as in rockwool, culture, sand culture, bag culture, and surface-irrigation gravel culture, and
b) Closed systems, where the nutrient solution is recycled and used continuously (requires addition of water and nutrients) as in all water culture methods, NFT, aeroponics, and gravel culture subirrigation.

2.1 Advantages and Disadvantages

The advantages and disadvantages of hydroponic culture over soil culture may be summarized as follows:

1. Balanced Nutrient Solution. The accurate control of the nutritional supply to the plant, and therefore growth control is achieved continuously. Since the nutrient solution, unlike soil, is homogeneous, it is relatively easy to sample, test, and readjust nutrient supply periodically by replenishing those elements that are lacking.

2. Locality Crops can be grown in localities where normal cultivation is difficult or impracticable, e.g., in arid areas of saline or shallow soils. This opens up new regions for settlement and provides established settlements in such regions with new sources of income.

3. Aeration. Adequate aeration of the root medium is achieved. Roots need air – or more precisely oxygen – for respiration, which influences nutrient absorption. A porous growth medium and controlled irrigation result in good air exchange in the root environment, and are the basis for achieving high yields in hydroponics. In liquid culture, aeration of the solution is achieved in one of three ways: (1) continuous aeration of the liquid (nutrient solution); (2) pumping air into the liquid; (3) continuous flowing of the aerated solution. Despite this, however, it is not always possible to achieve the required amount of aeration.

4. Water Consumption. The considerable reduction in the amount of water needed to produce a given crop is often the determining factor for adoption of soilless methods in arid regions. The quantity of water used per unit weight of crop produced is less than that required under conventional soil culture for hot and dry conditions; for instance, soil evaporation and seepage can be reduced.

5. Prevention of Root Diseases. Sterilization of the aggregate and containers is simple and inexpensive, and the next crop planting can be performed 24 h after sterilization.

Using surface irrigation, obviating reuse of the nutrient solution, the distribution of root pathogens is drastically minimized.

6. Saline Water. Because the nutrient solution can be adjusted and is constantly being replaced, water containing a relatively large amount of saline may be used.

7. Yield. It is easier to obtain an average of high yields in hydroponics compared to other growth methods, because a proper root environment and good nutrient conditions are more easily achieved.

A grower with limited experience or knowledge of the optimal methods required for soil cultivation, who has the interest and sufficient background to understand the principles of soilless culture, may well do better with the latter system than with conventional soil cultivation methods.

With some crops, denser planting becomes possible, which may result in a higher yield per unit area.

8. Labor.

a) Investment: the most important factor in consideration of soilless methods is the cost of construction of the installation. The more elaborate and automatic the system employed, the greater will be its efficiency, as well as its cost. Simple methods, such as utilizing beds constructed from local materials and the application of solutions manually through perforated pipes, may often be satisfactory but will not always provide all the possible advantages to be gained from soilless methods.

b) Knowledge required: in order to successfully utilize hydroponic methods, one must have some knowledge of plant physiological principles and sufficient information on the particular crop to be grown. It must, however, be noted that many of the claims made by enthusiasts of soilless culture, as well as by "organic material only" fanatics, are often based on emotional convictions, and have not been scientifically proven.

In addition to the aforementioned differences between soil and soilless culture, the following features are often distinguished by growers and scientists.

9. Roots and Rootlets. One of the most distinctive features distinguishing soil cultivation from soilless culture in morphological plant development is the absence of root hairs in most cultivated plants; in some cases, microscopic enlargements can be observed on the roots instead of root hairs. The total amount of root surface area is greatly reduced compared to that in soil culture.

The root volume, including attached gravel or other growth medium, is less than that found under soil conditions including attached soil. Moreover, a root depth of 18–20 cm is usually sufficient in hydroponics, whereas soil conditions require deeper growth sections.

With less roots, most of them are active and therefore more resistant to pathogens.

The main disadvantages of soilless culture are:

a) Relatively more expensive in initial investment, especially in gravel and water culture methods.
b) Environmental problems if (1) used nutrient solution spoils groundwater or (2) used growth material such as rockwool is not properly removed.
c) The small buffer capacity, especially in NFT and related methods.

3 Nutrients

3.1 The Elemental Composition of Plants

Although soils are composed largely of oxygen, silicon, and aluminum, plants growing in soil by no means reflect this composition. Carbon and oxygen, followed by hydrogen, make up the bulk of the dry weight of plants. These elements are obtained from carbon dioxide and water. More than 50 elements have been found in various plants, compared to the total of 92 known natural elements. Plants are able to absorb in their roots many of the existing elements available in solute form from the surrounding solution. Plants do have the partial ability to select the rate at which they absorb various ions, so that absorption is usually not in direct proportion to their concentration in the growth media. The species vary in their ability to select particular ions.

The criteria for essentiality are usually defined as follows.

First, the element must be essential for normal growth or reproduction, neither of which can occur in its complete absence.

Second, the requirement must be specific and cannot be replaced by substituting some other element.

Third, the element must act directly inside the plant and not simply cause some other element to be more readily available or simply antagonize a toxic effect of another element.

Regarding criterion one: it is recognized that if the amount of the element needed is less than 0.2 g per gram of dry plant material, it may be regarded as completely absent. This criterion was observed by using silicon, an element found in most soils, and silicon may be recognized as nonessential.

Regarding criterion two: several examples are known in which nonessential elements can partially satisfy the need for one that is essential; calcium can be satisfied by strontium, and chlorine by bromine, for example.

Regarding criterion three: for all its essential elements, essential metabolic processes, or at least stimulating processes, are present in the plants (Clarkson and Hanson 1980).

3.1.1 Elements Essential for All Plants

Sixteen elements are thought to be actually essential to the growth of most plants. Table 3.1 lists these 16 elements, the chemical forms most commonly available to plants, the concentration in the plant considered adequate, and the number of atoms of each element needed compared to molybdenum. The required elements are usually defined as macroelements or macronutrients, those required in relatively large quantities, and micronutrients, those required in small amounts. The macronutrients include carbon, hydrogen, oxygen, nitrogen, phosphorus, calcium, sulfur, potassium, and magnesium. The micronutrients include chlorine, iron, manganese, boron, zinc, copper, and molybdenum.

Some elements are essential in special plant groups, such as aluminum, silicon, cobalt, and sodium, but these elements are not essential for most of the plant groups, and are therefore considered nonessential elements.

The essential macroelements were discovered in the 19th century, whereas the micronutrient essentiality was determined later, as described in Table 3.2.

The essentiality of mineral elements of various plant groups is summarized in Table 3.3.

When an essential element is not in sufficient supply or proper form to adequately supply the plant, a nutrient deficiency occurs (see also Sects. 3.5 and 3.6). The concentration in the plant does not meet the plant's physiological requirements. Plants usually grow poorly and show

Table 3.1. Elements Essential For Most Higher Plants and Internal Concentrations Considered Adequate. (After Stout 1961, in Epstein 1972)

Element	Chemical symbol	Form available to plants	Atomic wt.	Concentration in dry tissue: ppm	Concentration in dry tissue: %	Relative no. of atoms compared to molybdenum
Molybdenum	Mo	$MoO_4^{=}$	95.95	0.1	0.00001	1
Copper	Cu	Cu^+, Cu^{2+}	63.54	6	0.0006	100
Zinc	Zn	Zn^{2+}	65.38	20	0.0020	300
Manganese	Mn	Mn^{2+}	54.94	50	0.0050	1000
Iron	Fe	Fe^{3+}, Fe^{2+}	55.85	100	0.010	2000
Boron	B	$BO_3^{=}$, $B_4O_7^{=}$	10.82	20	0.002	2000
Chlorine	Cl	Cl^-	35.46	100	0.010	3000
Sulfur	S	$SO_4^{=}$	32.07	1000	0.1	30000
Phosphorus	P	$H_2PO_4^-$, $HPO_4^{=}$	30.98	2000	0.2	60000
Magnesium	Mg	Mg^{2+}	24.32	2000	0.2	80000
Calcium	Ca	Ca^{2+}	40.08	5000	0.5	125000
Potassium	K	K^+	39.10	10000	1.0	250000
Nitrogen	N	NO_3^-, NH_4^+	14.01	15000	1.5	1000000
Oxygen	O	O_2, H_2O	16.00	450000	45	30000000
Carbon	C	CO_2	12.01	450000	45	35000000
Hydrogen	H	H_2O	1.01	60000	6	60000000

Table 3.2. Discovery of the essentiality of micronutrients for higher plants. (Marschner 1983)

Element	Year	Discovered by
Iron	1860	J. Sachs
Manganese	1922	J.S. McHague
Boron	1923	K. Warington
Zinc	1926	A.L. Sommer and C.B. Lipman
Copper	1931	C.B. Lipman and McKinney
Molybdenum	1938	D.I. Arnon and P.R. Stout
Chlorine	1954	T.C. Broyer et al.

Table 3.3. Essentiality of mineral elements for plants.[a] (Epstein 1972)

Elements	Higher plants	Algae	Fungi	Bacteria
N, P, S, K, Mg Fe, Mn, Zn, Cu	+	+	+	+
Ca	+	+	±	±
B	+	±	–	–
Cl	+	+	–	±
Na	±	±	–	±
Mo	+	+	+	±
Se	±	–	–	–
Si	±	±	–	–
Co	–	±	–	±
I	–	±	–	–
V	–	±	–	–

[a]Essential for the group: +; not known to be essential for the group: –; essential for some but not known to be generally essential for members of the group: ±.

visual signs of abnormality in color or structure. Under toxicity or excess supply of a nutrient element in the rooting media or nutrient solution, toxicity symptoms may occur in the plant. These excess supplies affect the plant's normal growth and development.

3.2 Macroelements

3.2.1 Nitrogen

Of all the elements added to the nutrient solution, nitrogen is used in the greatest amount. It mainly accumulates in the tissues of young plants. The high mobility of nitrogen in the plant enables it to be transferred from older to younger parts of the plant when the amount of nitrogen in the nutrient solution is insufficient for continued growth. Reutilization of nitrogen is the reason for deficiency symptoms sometimes appearing first in older leaves.

Sources. The plant takes up nitrogen either as nitrate (NO_3) or ammonium (NH_4) ions from the soil solution. In hydroponics, nitrogen is mostly supplied in the form of nitrates. Ammonium ion should only be used as a supplement, since too large an amount will affect growth adversely (Feigin et al. 1980).

Either potassium or calcium nitrate are the nitrogen salts generally used; they also supply either potassium or calcium, both important nutrient elements. Sodium nitrate is much cheaper than other nitrate salts, but is not generally used since sodium is not essential for growth. If the water itself contains appreciable amounts of sodium, sodium nitrate cannot be used, since this may raise the solution's sodium content to toxic levels. Ammonium nitrate contains both ammonium and nitrate ions, but since the ammonium content is high, it should not be used as the only nitrogen source; it may, however, be used with either potassium or calcium nitrate. All the nitrate salts have a basic (alkalizing) effect on nutrient solutions.

The cheapest nitrogen source is ammonium sulfate; this may exert an acidifying effect on nutrient solutions.

A good source of both ammonium and nitrate ions is calcium ammonium nitrate, which also contains large amounts of soluble calcium.

Uptake. Absorption of nitrate and ammonium ions is closely related to plant carbohydrate production and status. Absorbed nitrogen is useless to the plant until it is incorporated into an organic molecule; incorporation generally occurs in the root. A ready supply of carbohydrates and an energy source are needed for nitrogen incorporation. When days are long and nights are cool, sugar production is favored and the plant can incorporate greater amounts of nitrogen. After uptake, nitrates are reduced to the ammonium ion and then incorporated as such. Ammonium ion uptake is, therefore, followed much more quickly by its incorporation. The incorporation of nitrates is retarded when the plant is short of sugar.

Deficiency Symptoms. The most general symptom of nitrogen deficiency is restricted growth of both tops and roots. Shoots are short. Growth is perpendicular and spindly. Leaves are small and usually lighter green. An early symptom is paling of very young leaves. In later stages older leaves develop red, orange, and yellow spots; this tinting then extends to younger leaves. The last stage consists of defoliation, dormancy, or death of young buds, reduced blossoming, and low fruit yields.

Tomato: fruit production is reduced. Fruits are light green and small. Root systems are brown, larger in relation to the tops than in healthy plants.

Cucumber: stunted growth and yellow-green leaves are the first symptoms. The leaves later become completely yellow and stems hard and woody. Fruits are light in color.

Radish: plants are small. Older leaves are yellow with purplish lower veins. Roots are small and woody.

Floral crops: leaves are light green. Older leaves yellow and dry. Growth is woody.

Toxicity. In the case of nitrates this is seldom a problem. Excess ammonium, however, causes too rapid development of the plant's green parts and too rapid stem elongation. There is a lack of proportion between stems and leaves. In hard-leaved plants the leaves droop. High nitrogen content is conducive to blossom-end rot in tomato.

Excess of nitrogen, especially in the form of ammonium, causes nitrite development in the nutrient solution, with possibly toxic effects on plant growth.

3.2.2 Phosphorus

The levels of this element are higher in fruiting than in vegetative parts of plants. It is relatively mobile in plants, though it mostly accumulates in younger, growing tissues.

Sources. The only form of phosphorus available to plants is the oxidized phosphate ion (PO_4). Any soluble salt containing this phosphate ion, e.g., monopotassium or monoammonium phosphate, some form of superphosphate, or phosphoric acid, can therefore be added to nutrient solutions. Although potassium or ammonium phosphates are much more soluble than the superphosphates, they are generally too expensive for commercial use. The salts usually used in commercial hydroponics are single, double, or triple superphosphate. These differ in their phosphorus content – about 16% in single, 32% in double, and 43% in triple superphosphate. These salts all contain large amounts of calcium.

Often, weak phosphoric acid may be the most economic phosphorus source. In this case, a small amount of potassium hydroxide should be added to correct the solution's acidity. Potassium hydrogen phosphate is generally used in experimental work.

Uptake. Only dissolved phosphate is available to plants. Thus, in preparing nutrient solutions great care must be exercised to prevent precipitation of phosphate salts. Alkaline aggregates such as calcareous gravel will precipitate most of the phosphate as calcium or magnesium salts.

Deficiency Symptoms. Phosphorus-deficient plants are small, woody, blue-green in color, and the leaf veins have a purplish tinge; sometimes the entire lower leaf surface may become purplish. Fruit, if produced, is small. Roots are generally stunted. Leaves are small and the older ones drop. Fewer shoots are formed and lateral buds die. Budding and blossoming are delayed. In extreme cases the leaf margin turns brown.

Toxicity. Excess phosphorus precipitates the nutrient solution's iron ions, especially at a low pH. Since this causes iron deficiency, low levels of phosphorus should be maintained. The interaction between phosphorus and iron is the most common "toxic" phosphorus effect under soilless culture.

3.2.3 Potassium

Relatively large quantities of potassium are found in growing points, leaves, and in cell sap.

Sources. Potassium sulfate is the best source for commercial use. Potassium nitrate or phosphate are good potassium sources that also supply other essential nutrients; however, owing to their expense, they are generally used less in commercial hydroponics. Potassium chloride raised the chloride content of nutrient solutions and is not always recommended. Potassium hydrogen phosphate is expensive and contains insufficient potassium. However, in experimental work it is used with other potassium salts.

Deficiency Symptoms. The first signs of potassium deficiency usually are browning of the tips of older leaves and marginal scorching. Yellow-brown spots appear near the margin and the leaves may curl upward. Internodes are short, roots badly developed, and leaf margins cracked. Other deficiency symptoms, especially when extreme, sometimes simulate calcium deficiency, chloride, toxicity, salt burn, or dry-wind injury.

Tomato: plants are woody. Leaves become bluish green, yellow intervenous spots appear, leaf margins curl up and, in more extreme cases, become brown. Fruit ripens unevenly and has a patchy appearance.

Cucumber: in the first stages, leaf tips are light green, later the entire plant turns yellow-brown and then dies. The fruit tip is shrunken and bears black spots.

Radish: a dark green spot appears in the middle of the leaf, leaves curl up and the margins grow brown and dry.

Toxicity. Excess potassium causes enlarged internodes, light green leaves and, later, stunted growth and brown spots. Toxicity symptoms sometimes simulate those of calcium, and magnesium deficiency. Toxic effects may sometimes be lessened by adding more calcium to the nutrient solution.

3.2.4 Magnesium

Owing to the high mobility of magnesium, the first symptoms of its deficiency always appear in older leaves – as a chlorotic mottling. Leaf margins may remain green, but sometime curl up. If deficiency continues, chlorosis extends from older to younger leaves. Deficiency occurring in later growth stages will not affect fruit yields.

Toxicity. With excess magnesium, leaves become dark green, small and younger leaves curl up. Toxic effects may be lessened by balancing with high calcium concentrations.

3.2.5 Sulfur

Generally thought to be immobile.

Sources. Sulfur is usually supplied and taken up as the sulfate ion. Ammonium sulfate (for nitrogen), magnesium sulfate (magnesium), calcium sulfate (calcium), potassium sulfate (potassium), and superphosphate (phosphorus) are all rich in available sulfate ion, so that no sulfur salts need be added to nutrient solutions.

Deficiency Symptoms. Sulfur deficiency is rare in hydroponics, since the plant requires small amounts and relatively large amounts are supplied as sulfates. Deficiency symptoms usually develop gradually, and in the first stages resemble those of nitrogen deficiency. Plants are stunted and woody, stems slender. Chlorosis appears in older leaves, internodes are longer than normal, and older leaves may be thicker than younger ones.

Toxicity. A very late sulfur excess is needed for toxicity to become evident; it is thus rare. The symptoms are bluish green, small, curled leaves. High sulfate concentrations may limit plant calcium uptake.

3.2.6 Calcium

Usually considered immobile. If calcium is insufficient, no transfer from older to younger areas occurs.

Sources. The best source is calcium nitrate. This also contains nitrate but is usually expensive for commercial use. The cheapest sources are gypsum (hydrated calcium sulfate) and calcium chloride. The latter is generally not recommended, because of its high chloride content. Superphosphate – a form of calcium phosphate – contains much calcium, but not all of it is in soluble, i.e., available, form. Tap and saline waters in arid regions may contain sufficient calcium ions, requiring little or no additional calcium.

Deficiency Symptoms. Calcium deficiency first affects young leaves and root and stem growing points. Young leaf margins curl back; brown spots and a thin chlorotic, marginal band appear. The leaves also lose their regular marginal shape. Roots are poorly developed, root growing points die quite early, especially in root crops and tomatoes. In leaf crops, stems collapse near the growing point.

Tomato is especially affected by calcium deficiency. First the flower trusses die, roots turn brown and blossom-end rot (an apical black spot on the fruit) appears.

Calcium deficiency may appear under high saline conditions and often, if much sodium is present, even with solutions containing large amounts of calcium.

Toxicity. Calcium excess caused chlorosis and defoliation similar to that in magnesium and iron deficiencies. It interferes with potassium uptake, and plants may show potassium deficiency symptoms even with solutions containing sufficient potassium. Iron deficiencies may also appear if calcium levels are high. Chlorine accumulation is more pronounced with excess of calcium.

3.3 Microelements

The ranges for use of some microelements in nutrient solutions are narrow. An example is boron, a few ppm marking the difference between deficiency and toxicity. Plants can absorb a certain quantity of nutrients through their leaves. Sprays are thus often used to correct microelement deficiencies.

3.3.1 Iron

Sources. Iron is usually supplied as ferrous sulfate. If the solution's pH is over 7, ferrous ions are oxidized to ferric ions, and then an insoluble ferric sulfate precipitate appears. This form of iron is not available to plants. Iron may be supplied as ferric chloride or as the soluble organic salts – iron tartrate, iron citrate, or ferric citrate.

A method to be employed with caution to ensure persistent availability of iron to the plant is use of an iron chelate such as Sequestrene (Fe-EDTA). Chelates also prevent phosphate precipitation of iron, necessitating addition of iron to nutrient solutions only once a month instead of every 4–5 days, as is necessary with other forms of iron. Too much chelate may be deleterious since it may remove copper and zinc from solutions and cause deficiencies of other such heavy metals. Available chelates are Sequestrene 138 (6% iron) and Sequestrene 33 (10% iron).

Deficiency Symptoms. The first symptom of iron deficiency is chlorosis in the younger tissues. The amount of iron in the plant depends on the amounts of manganese, potassium, and other elements in the tissue; deficiency symptoms will therefore appear even if the solution contains sufficient iron but high concentrations of manganese, zinc, copper, calcium, or phosphorus. They will also appear if solutions are deficient in calcium, magnesium, or potassium. At later stages chlorosis appears in older leaves.

In tomatoes, leaves are mottled brown and new leaves are chlorotic; this is most pronounced near the midrib and at the base of leaflets. Tissue of strawberry leaves has chlorotic areas while veins remain green.

3.3.2 Boron

Sources. Boron is supplied either as boric acid or borax, and the borate ion (BO_3) is taken up.

Deficiency Symptoms. Boron deficiency appears first at plant growing points, which die. Leaves are later slightly mottled and flowers and stems are roughened. Buds die and leaves then become yellow-grown and curl inward.

In lettuce, the young leaves are weak; they grow brown and die. The growing point dies and older leaves become deformed. In tomatoes, the growing point dies, the plants becoming bushy, with brittle stems and petioles. In legumes leaves become chlorotic.

Toxicity. The optimal range for boron is the narrowest for all the nutrient elements. Toxic amounts of boron may be present in saline waters. Excess boron caused marginal and intervenous leaf scorch and chlorosis. In tomatoes, transparent areas appear and later turn brown. Excess boron may delay the appearance of calcium deficiency symptoms, as it may to some extent replace calcium's physiological functions.

3.3.3 Manganese

Sources. Manganese is generally supplied as manganese sulfate or chloride.

Deficiency Symptoms. Manganese deficiency symptoms first appear in younger leaves as a tissue necrosis (leaf transparency). This occurs especially in leaves farther away from the main stem, in which the veins retain their color.

In tomatoes and cucumbers, the first symptoms are brown spots on the younger leaves that may spread to older parts of the plant, the whole plant becoming pale green. In lettuce, leaves are small and show intervenous necrosis.

Toxicity. Toxic effects appear only with relatively large amounts of manganese. Lettuce is affected at about 10 ppm, tomatoes at a much higher level. Exact toxic levels depend on total salt content of solution. General symptoms of excess manganese are curling-in of the leaves and death of the growing point. The leaf margin usually also becomes spotted and scorched.

3.3.4 Zinc

Sources. Zinc is usually supplied as zinc sulfate or chloride.

Deficiency Symptoms. Plant zinc accumulation and light quantity are apparently related; zinc deficiency may be more frequent in summer than in winter.

Insufficient zinc causes leaves to be abnormally small, mottled with chlorotic spots, and curled upward. Symptoms first appear in younger leaves.

Toxicity. Important sources of zinc at possibly toxic levels are galvanized (zinc-coated) iron pipes and equipment, from which zinc is dissolved and carried into the nutrient solution. Under saline conditions, zinc levels that are normally toxic may have no deleterious effects.

3.3.5 Copper

Sources. Copper is usually supplied as copper sulfate or chloride.

Deficiency Symptoms. Copper deficiency generally causes chlorosis, especially in the younger plant parts, and stunted growth. Leaves are green-blue and margins are sometimes burned.

Toxicity. Copper and copper compounds are well known as fungicides. Copper is toxic in high concentrations but this is seldom in practice. With saline waters, phosphorous addition also reduces the toxic effects of the nutrient solution's high copper levels. Sensitivity to copper varies greatly among different plant species. Lettuce is sensitive; vines, pineapples, and peas are tolerant to relatively large amounts.

In tomatoes, excess copper may cause symptoms similar to those of iron deficiency. Large amounts of copper accumulate in the roots and darken them.

3.3.6 Chloride

A nutrient solution containing about 10–20 ppm of chloride ions is required for optimal growth.

Sources. Most natural waters contain sufficient chlorine for normal growth. Chlorides are generally added with the use of technical-grade chemicals. Deficiency is extremely rare.

Toxicity. One should distinguish between those toxic effects due to excess of chloride, high sodium levels, and high osmotic pressures caused by high salt concentration. Saline effects are not solely those due to high chloride content. With excess chloride, leaves are small and dull green, and the plant becomes woody. Earlier flowering in tomatoes has been ascribed to excess chloride in the solution.

3.3.7 Molybdenum

Sources. Molybdenum is required in such minute quantities that the amounts present as impurity in the nutrient chemicals are generally sufficient.

Bicarbonate. High levels of this ion may be toxic, plant species varying in their tolerance to it. It affects plant nutrient uptake and metabolism. The general toxic symptom is chlorosis. An excess may occur under saline conditions.

Bromine. Bromine has not been found to be essential for normal plant growth. If, however, chlorine is present in small amounts, bromine may partly replace it.

3.4 Nonrequired Elements

Certain plants demonstrate a definite need for some elements which most plants do not require at all, and which in some cases are actually detrimental to the development of some plant species.

Sodium. Sodium is not known to be generally required by green plants. However, certain halophytes, plants indigenous to saline soils, not only tolerate the high concentrations of salt in such soils but actually require sodium, such as *Atriplex*. Sodium is found in most water sources, and especially in saline water. Small amounts of sodium are well tolerated by most culture plants.

Silicon. This is not required by most culture plants. It is required in rice. Silicon is found in most water sources and in many growth media. Small amounts of silicon are well tolerated.

Cobalt. Required in plants which are hosts to nitrogen-fixing bacteria contained in root nodules, such as legumes. These requirements occur in several soil conditions. Legumes can be grown under soilless conditions without cobalt.

Aluminum. Like silicon, this is found in most plants. A few milligrams of aluminum stimulate the growth of some species, but it is generally regarded as a nonrequired element.

3.5 Deficiency Symptom Considerations

The metabolic disarrangements brought about by deficiencies of essential nutrients eventually manifest themselves in visible abnormalities. The overall growth and development of the plant may be affected, there may be characteristic macroscopic symptoms, and there may be changes in the appearance of cellular structures that can be studied with the microscope or electron microscope.

The overall appearance of the plant and localized specific symptoms are an important aid to knowledgeable observers in identifying deficiencies. The symptoms of iron deficiency, for example, are fairly similar in many crop species, and sometimes are all that is needed to pinpoint a deficiency of this element. Nevertheless, visible symptoms should be regarded as just one of several kinds of evidence of deficiency of a given element.

Symptoms of a deficiency of a certain element may differ so greatly in different crops that knowledge of the deficiency syndrome in one species affords little aid in identifying the same deficiency in another species. Similar or identical symptoms may result from deficiencies of different elements. For example, both nitrogen and sulfur deficiencies cause a general chlorosis, and even experienced observers may often be unable to distinguish between deficiencies of these elements on the basis of observation alone. Deficiencies may be severe enough to reduce yields and impair the quality of the crop, while the visible symptoms are still too slight to be readily spotted.

Multiple deficiencies are difficult to determine. A plant may have symptoms of deficiencies of two or more elements, making identification of the deficiencies more difficult than when they occur singly. Conditions that are not deficiencies at all may engender symptoms which might be mistaken for deficiency symptoms. Such "simulated" deficiency symptoms may be caused by many factors ranging from elemental toxicities to bacterial diseases and virus infections.

In the appendix a general identification chart is given to provide the grower with alternative possibilities of observed deficiency symptoms (Table 17a, b in Appendix).

3.6 Interaction Between Nutrients

Interaction between nutrients in higher plants occurs when the supply of one nutrient affects the absorption, distribution, or function of another nutrient. Thus, depending on nutrient supply, interactions between nutrients can induce either deficiencies or toxicities and can modify growth response. Interactions occur between ions because of the capability of some ions to form a chemical bond which will produce precipitates or complexes with organic ligands. Interaction may occur when ions with similar chemical properties compete for sites of absorbtion (such as the competition for absorbtion, transport, and function in cell walls). Table 3.4 shows the interactions occurring between elements.

Table 3.4. Interactions occurring between elements which affect the absorption and utilization of nutrients by plants

	NH_4^+	K^+	Na^+	Ca^{2+}	Mg^{2+}	Mn^{2+}	Cu^{2+}	Zn^{2+}	Fe^{2+}	Al^{3-}		Cl^-	NO_3^-	H_2PO_4	SeO_4^{2-}	SO_4^{2-}	MoO_4^{2}	H_3BO_3	H_4SO_4
NH_4^+																			
K^+	+																		
Na^+		+																	
Ca^{2+}		+	+																
Mg^{2+}		+		+															
Mn^{2-}				+															
Cu^{2-}																			
Zn^{2-}				+	+		+												
Fe^{3+}/Fe^{2+}				+		+													
Al^{3+}				+			+												
Cl^-				+															
NO_3^-												+							
$H_2PO_4^-/HPO_4^{2-}$				+			+	+	I	I		O	O						
SrO_4^{2-}																			
SO_4^{2-}												O	O		+				
MnO_4^{2-}							+		+					+		+			
H_2BO_3				+										+					
H_4SiO_4						+								+					

+, interaction recorded; °, no interaction observed after investigation; no satisfactory data for other combinations.

3.7 Methods of Expressing Solution Concentration

Several forms of expressing the concentration of salts in a solution are given below:

Milligrams per liter (mg/l): the weight of salts in milligrams contained in a liter of solution.

If one gram (1000 milligrams) of table salt is dissolved in a liter of water the solution is said to contain 1000 milligrams of salt per liter – abbreviated to 1000 mg/l.

Parts per million (ppm): a milligram is one millionth part of a kilogram; since a liter of water weighs one kilogram, 1 mg/l is the same as 1 mg/kg or 1 part per million (1 ppm).

Moles and millimoles per liter (M/l and mM/l): the molecular weight of salt (or any other chemical compound) is obtained by simple addition of the atomic weights (available from standard tables) of its constituent elements. In the case of $NaNO_3$ (sodium nitrate), which consists of one atom of sodium (atomic weight = 23), one atom of nitrogen (atomic weight = 14), and three atoms of oxygen (atomic weight = 16), the molecular weight = $23 + 14 + 3 \times 16 = 85$. The molecular weight expressed in grams is termed a mole (gram molecule). A liter of solution containing 85 g of $NaNO_3$ has one mole of this salt per liter (1 M/l); a liter containing 42.5 g had 0.5 mole, 17 g 0.2 mole etc. Since one mole is equivalent to 1000 millimoles, the one-mole solution contains 1000 millimoles, the 0.5 mole 500 millimoles, etc.

Percent (%) w/v: this is part of salt per 100 parts of solution. If one liter of a solution contains 50 g salt, then it is said to be a 5 percent (5%) solution, on a weight on volume (w/v) base.

There are various ways of expressing the amount of a specific ion in a solution. The amount of nitrogen in a solution may be expressed in terms of elemental nitrogen (N), nitrogen as a nitrate in (NO_3^-) or as nitrite ion (NO_2^-). The amount of phosphorus may be expressed as elemental phosphorus (P), as phosphate ion (PO_4^{-3}), or as phosphorus pentoxide (P_2O_5). For example, 14 parts of nitrogen expressed as N is equal to 62 parts expressed an NO_3^-.

3.7.1 Adding Nutrient Salts

The method of making up the solution depends on the amount and the salts used. Three types of nutrient solutions are generally used.

1. Single (normal) concentration: 1 l of solution contains the amount of salts giving optimal plant growth.
2. Double twice normal concentration: 1 l of solution contains double the amount of salts in a normal concentration.

3. Half-concentration (half-normal): the same amount of salts as in a normal solution is made up to 2 l of solution.

In winter, evaporation and transpiration water losses are decreased and double concentration solutions may be used. In summer, evaporation and transpiration water losses are high and half-concentration solutions may be used. Triple or greater concentration solutions may be made if needed.

Macronutrient salts may be weighed on a balance with a 3–5% accuracy, i.e., if 100 g are needed, an amount ranging from 98–102 g will be obtained, and for 10 kg, 9.8–10.2 kg. Greater balance error is unadvisable.

There can be a significant difference in the cost, purity, and solubility of the chemicals comprising a nutrient solution, depending on the grade (pure, technical, food, or fertilizer) used. Smaller operators often buy ready-mixed nutrient formulations; only water need be added to prepare the nutrient solution. Larger facilities prepare their own solutions to standard or slightly modified formulas. The commonly used weight factors in grams required to make 1000 liters of a 1-ppm solution are given in Appendix Tables 2–9. It is necessary only to multiply the factor for a chemical by the number of ppm desired in the formula to obtain the number of grams to be used per kiloliter.

The local availability and cost of fertilizers often determine what components are used. Using these tables, one can easily prepare virtually any formula; if a readily available chemical is not listed, its weight factor can be calculated from the atomic weight.

The preparation of typical nutrient solutions for tomato and cucumber culture hydroponic systems is outlined in Table 3.5. A micronutrient solution designed to supplement these basic solutions is described in Tables 3.5 and 3.6 and in Appendix Tables 3, 4, 6, and 8. For larger hydroponic systems, chemicals are weighed out individually to an accuracy of ± 5% (small deviations generally have no apparent effect on plant growth) and arranged near the mixing tanks in a manner that precludes workers from double-weighing any component. The chemicals are simply added to the tanks and stirred vigorously.

Table 3.5. Composition of hydroponic nutrient solutions. (Resh 1983)

	Element (mg/kg)											
Reference	N	P	K	Ca	Mg	S	Fe	Mn	Cu	Zn	B	Mo
Hoagland and Arnon (1950)	210	31	234	160	48	64	0.6	0.5	0.02	0.05	0.5	0.01
Schwarz (1968)	126	93	312	124	43	160	–	–	–	–	–	–
Resh (1983)	175	65	400	197	44	197	2	0.5	0.03	0.05	0.5	0.02
Verwer (1976)	173	39	280	170	25	103	1.7	1.1	0.017	0.25	0.35	0.058
Graves (1983)	175	50	400	225	50	–	3	1	0.1	0.1	0.4	0.05

Table 3.6. Nutrient solution concentrations. (Epstein 1972)

Compound	Molecular w	Concentration of stock solution (M)	Concentration of stock solution (g/l)	Volume of stock solution per liter of final solution (ml)	Element	Final concentration of element (μM)	Final concentration of element (ppm)
		Macronutrients					
KNO_3	101.10	1.00	101.10	6.0	N	16000	224
$Ca(NO_3)_2 \cdot 4H_2O$	236.16	1.00	236.16	4.0	K	6000	235
$NH_4\ H_2\ PO_4$	115.08	1.00	115.08	2.0	Ca	4000	160
$MgSO_4 \cdot 7H_2O$	246.49	1.00	246.49	1.0	P	2000	62
					S	1000	32
					Mg	1000	24

Compound[a]	Molecular w	Concentration of stock solution (mM)	Concentration of stock solution (g/l)	Volume of stock solution per liter of final solution (ml)	Element	Final concentration of element (μM)	Final concentration of element (ppm)
		Micronutrients					
KCl	74.55	50	3.728	1.0 (KCl through H_2MoO_4)	Cl	50	1.77
$H_3\ BO_3$	61.84	25	1.546		B	25	0.27
$MnSO_4 \cdot H_2O$	169.01	2.0	0.338		Mn	2.0	0.11
$ZnSO_4 \cdot 7H_2O$	287.55	2.0	0.575		Zn	2.0	0.131
$CuSO_4 \cdot 5H_2O$	249.71	0.5	0.125		Cu	0.5	0.032
$H_2\ MoO_4$ (85% MoO_3)	161.97	0.5	0.081		Mo	0.5	0.05
Fe-EDTA[b]	346.08	20	6.922	1.0	Fe	20	1.12

[a]A combined stock solution is made up containing all micronutrients except iron.
[b]Ferrous dihydrogen ethylenediamine tetraacetic acid.

3.8 Composition of the Nutrient Solution

Reference to the available literature reveals that there may be considerable diversity in the composition of the nutrient solution used. To date, a total of approximately 300 formulas have been published, many of which have been recommended for the growth of specific crops, e.g., tomatoes, carnations, etc.

Physiological experiments on nutrition have been performed to investigate the composition of a wide range of nutrient solutions and their influence on the development and behavior of plants. For experiments of this type, a knowledge of the problems inherent in the preparation of a desired nutrient solution is a prerequisite, since the actual composition may not always correspond to that given by the formula (Hewitt and Smith 1975; Marschner 1983).

The observation that a crop grows well in one particular nutrient solution does not prove that there is not a more favorable one for that crop.

Since the plant itself exercises certain selective properties, the question may be raised as to the extent to which the composition of the nutrient solution is important for growth and development. The chemical composition of a nutrient solution comprises (1) the concentrations of the component ions, (2) the total ionic concentration, and (3) pH. Experiments with flowing cultures, in which fresh nutrient solution is supplied continuously to plants, have shown that individual ions are taken up even at very low concentrations, but it is probable that, for every ion, there is a certain minimum concentration below which uptake is no longer possible at the required rate. At the other extreme, luxury consumption may lead to internal toxicity. The point at issue is whether there is an optimum concentration of certain ions in a nutrient solution under given environmental conditions, or whether their relative proportions and not absolute concentrations are the determining factors (Marschner 1983).

Some nutrient concentrations and nutrient solutions are given in Tables 3.5–3.8.

3.8.1 Changes in the Composition of the Nutrient Solution

An important difference between soilless culture and cropping in soil rests upon the fact that the buffering capacity in hydroponic systems is very restricted. Therefore factors which influence the composition of the nutrient solution, such as nutrient and water uptake, for instance, can affect the conditions of the nutrient medium much more in hydroponic systems than in soil.

In principle, the apparent effect which the composition of the nutrient solution has on plants depends on the relation between water uptake and nutrient uptake. If the uptake of water and nutrients occurs in the same proportion as prepared in the nutrient solution, the composition of the nutrient solution will not change. A comparatively larger uptake of water will lead to an increase in concentration of nutrient elements and this is often the case. If the nutrient uptake is greater than the water uptake, the nutrient concentration will decline, a situation which does not usually occur. Other factors which influence the composition of the nutrient solution are the substrate, the walls of the culture vessels, and the water quality.

Although many failures are erroneously blamed on the composition of the nutrient solution, difficulties are often due to imperfect physical circumstances in the root environment.

Table 3.7. Preparation of nutrient solutions for tomato and cucumber culture in closed or open hydroponic systems[a]. (Jensen and Collins 1985)

		Tomato		Cucumber	
		Soln. A	Soln. B	Soln. C[b]	Soln. D
Chemical compound (fertilizer grade)	Principal element supplied	Seedlings to first fruit set (g/1000 l)	Fruit set to harvesting (g/1000 l)	Seedlings to first fruit set (g/1000 l)	Fruit set to harvesting (g/1000 l)
Magnesium sulfate $MgSO_4 \cdot 7H_2O$ (Epsom salt grade)	Mg	500	500	500	500
Monopotassium phosphate KH_2PO_4 (0–22.5–28.0)	K, P	270	270	270	270
Potassium nitrate KNO_3 (13.75–0–36.9)	K, N	200	200	200	200
Potassium sulfate[c] K_2SO_4 (0–0–43.3)	K	100	100	–	–
Calcium nitrate[d] $Ca(N)_3)_2$ (15.5–0–0)	N, Ca	500	680	680	1357
Chelated iron[e] FE 330	Fe	25	25	25	25
Micronutrients		150 ml	150 ml	150 ml	150 ml

[a]Final nutrient concentrations in mg/kg: Soln. A – Mg(50), K(199), P(62), N(113), Ca(122), and Fe(2.5); Soln. B – Mg(50), K(199), P(62), N(144), Ca(165), and Fe(2.5); Soln. C – Mg(50), K(154), P(62), N(144), Ca(165), and Fe(2.5); Soln. D – Mg(50), K(154), P(62), N(260), Ca(330), and Fe(2.5). All solutions are supplemented with micronutrients.
[b]Solution C can be used for other vegetable crops; adjust N levels to 200 ppm for leafy vegetables such as lettuce.
[c]The use of potassium sulfate is optional.
[d]Adjust N levels to 200 ppm for leafy vegetables such as lettuce.
[e]Up to 50 g/1000 l may be necessary if a calcareous growing medium is used.

Table 3.8. Preparation of micronutrient solution for tomato and cucumber culture. (Jensen and Collins 1985)

Chemical compound[a]	Element supplied	Grams to use[b]
Boric acid (H_3BO_3)	B	7.50
Manganous chloride ($MnCl_2 \cdot 4H_2O$)	Mn	6.75
Cupric chloride ($CuCl_2 \cdot 2H_2O$)	Cu	0.37
Molybdenum trioxide (MoO_3)	Mo	0.15
Zinc sulfate ($ZnSo_4 \cdot 7H_2O$)	Zn	1.18

[a]Final nutrient concentrations in mg/kg: B(0.44), Mn(0.62), Cu(0.05), Mo(0.03), and Zn(0.09).
[b]Add water to mixture of micronutrients to make 450 ml of stock solution. Use 150 ml of this micronutrient solution with each 1000 l of nutrient solution

Steiner (1980) observed that the anion nitrate, phosphate, and sulfate uptake by lettuce and tomato plants is in a very specific mutual ratio, irrespective of a widely divergent mutual anion and cation ratio, while the mutual cation calcium, magnesium, and potassium uptake ratio is in a less extreme but still mutual ratio. This means that if there are relatively small deviations in the chemical composition of the nutrient solution, the plants themselves have the "mechanism" to select the ions in a mutual ratio favorable to their growth and development (Adams 1980; Clarkson and Hanson 1980; Sonneveld 1981; Robson and Pitman 1983; Van Goor et al. 1988).

The various osmotic pressures and concentration ratios of N, P, S, K, Ca, and Mg in the nutrient solution have been tested in tomato plants at various phases of growth as illustrated in Fig. 3.1 and explained in the following paragraphs.

The osmotic pressure itself has a rather strong influence on the mutual ratio of ion uptake. This influence is independent of the growth phase. At a higher osmotic pressure plants absorbed much more potassium, mainly at the cost of calcium. A higher osmotic pressure caused a strong preference for phosphate and a rather small preference for nitrate, both at the cost of the sulfate uptake.

The conclusions are:

1. A tomato plant has a strong selective capacity for ions.
2. The direction of the selection depends mainly on the growth phase of the plant, from which follows that we may use the same mutual ration between the ions in the nutrient solution during the entire plant life.

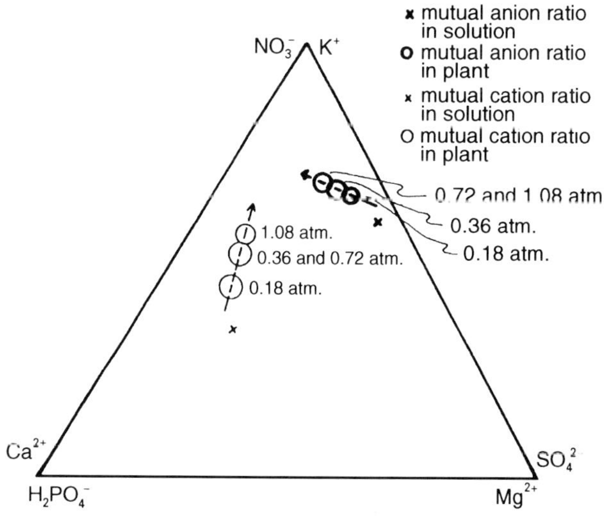

Fig. 3.1. The influence of the osmotic pressure of the nutrient solution on the mutual ratio of ion uptake in equivalents

3. There are indications that the direction of the selection is not specifically influenced by light intensity or air temperature, from which follows that we may use the same mutual ion ratio during summer and winter.
4. The osmotic pressure of the nutrient solution has a rather strong influence on the direction of selection for ions.

Steiner raises the question as to whether it is the relative concentration of elements among each other that determines uptake or the absolute amount. He suggests that there must be a minimum concentration below which uptake is no longer possible and above which luxury consumption occurs leading to internal toxicity. However, within this range, there also must be relative relationships that determine uptake and, therefore, the composition of the nutrient solution must be in a particular balance to satisfy the plant requirement for essential elements.

Steiner's aim was to determine how a particular nutrient solution could be prepared which satisfies given requirements as to:

1. relative cation ratios,
2. relative anion ratios,
3. total ionic concentration, and
4. pH.

Concerning himself with three major acions and three major cations, Steiner established the equivalent ratio in percent as follows (example):

NO_3^-	$:H_2\ PO_4^-$	$:SO_4^{2-}$
50	:7	:43
K^+	$:Ca^{2+}$	$:Mg^{2+}$
50	:30	:20

Using five different source chemicals and aliquots to establish the desired ratio or ions and not to exceed a total of 30 mg ions/l, Steiner's formula was:

Chemical	Normality (N)	ml/10 l
KH_2PO_4	1	8.22
$Ca(NO_3)_24H_2O$	10	1.644
$MgSO_47H_2O$	2	8.22
KNO_3	1	115.07
K_2SO_4	1	8.22

Preparing various nutrient solutions with the objective of maintaining the desired ratio of ions and pH, it became evident to Steiner that if the total ionic concentration was raised above 30 mg ions per liter and the pH above 6.5, only a very few combinations were possible in order to avoid problems due to precipitation.

Steiner's study reveals the possibility of preparing nutrient solutions that have specific ratios of ions to each other, a set total ion concentration and pH. He sets forth an interesting approach to nutrient solution formula development that bears further study for the inclusion of the other essential elements, forms, such as NO_3^- and NH_4-N, and technique for use.

3.9 Mixed Fertilizers

Offered by fertilizer companies, mixed fertilizers relatively inexpensive source for preparation of a nutrient solution. The contents of these fertilizers were generally described as percentages of the ions in the mixture, e.g., 20–5–15 mixture means that 20% of the fertilizer is N, mostly in the form of NH_4, 5% is P, mostly in the form of K_2O.

The amount of dissolved ions in water from these fertilizers, and therefore generally accepted to be in available form for plant intake, is often not equivalent to the percentage stated by the firms. Analysis of a dissolved sample of this fertilizer will determine the actual relationship between the ions.

As most of these fertilizers contain their N in ammonium form, these mixtures are of only limited use. Certain fertilizer mixtures contain sufficient amounts of microelements, and these contents are stated by the company.

Complete nutrient solution dry mixtures are relatively expensive. These mixtures are usually supplied in two combinations, and, when dissolved in a solution, represent a sufficiently good balance of nutrients.

3.9.1 Chelating Agents

The micronutrient cations iron, manganese, zinc, and copper are partly insoluble in nutrient solutions when provided as common inorganic salts. This insolubility is especially marked if the pH is held above 6, since cations react with hydroxyl ions, precipitating out the hydrous metal oxides, as shown in the following example, where the ferric form of iron is precipitated as a hydrated oxide (rust).

$$2Fe^{+3} = 6OH^- \rightarrow 2Fe(OH)_3 \rightarrow Fe_2O_3 \cdot 3H_2O.$$

Because of these undesirable reactions, some metals – especially iron – are often added as metal chelates (from the Greek, clawlike). A chelate or complex ion is the rather stable product formed when certain atoms in an organic chelating agent (ligand) donate electrons to the metal cation. One of the best-known synthetic ligands provides both carboxyl groups and nitrogen atoms. This is ethylene-diaminetetraacetic acid, abbreviated EDTA.

A good chelating agent forms stable chelates with the micronutrient ions and not with more abundant competitive cations such as calcium or magnesium. EDTA has a high affinity for calcium ions and is thus a poor chelating agent in high calcareous and magnesium nutrient solutions, as may occur under some saline conditions.

EDTA chelate, or Versenate, is commonly used and is included in most nutrient solution mixtures (Fischer 1984).

3.10 pH

The pH affects salt absorption in various ways. At a low pH, the hydrogen ions usually decrease the absorption of cations, while anion absorption may be stimulated. The apparent reason for this is that hydrogen ions compete with cations for uptake sites, while hydroxyl or bicarbonate ions present at a higher pH compete with anions such as nitrate, chloride, and phosphate. The pH also strongly influences the uptake of phosphate because it influences the ionic charge. The predominant ionic forms are $H_2PO_4^-$ at low pH values, an equal mixture of $H_2PO_4^-$ and $HPO_4^=$ at pH 6.8, $HPO_4^=$ at higher pH values. Monovalent forms are usually absorbed more readily than divalent, and divalent faster than trivalent. This is certainly true for phosphate, and the inability of some plants to grow at a high pH may be due to a limited rate of phosphorus absorption. The poor availability of iron and certain other micronutrients is probably also involved in the slow growth of plants at a high pH (Islam et al. 1980).

3.11 Mobile and Immobile Nutrients

A number of nutrients such as N, P, and K move readily to other plant organs from leaves in which they have accumulated. Indeed, their movement to young growing organs may continue to the detriment of the old leaves in which their concentrations may fall to deficient levels.

Plants which have accumulated excess concentrations in their leaves may continue to grow unchecked even when they have no external supply of these nutrients. Moreover, deficiencies do not develop until the total amount of nutrient in the plant as a whole becomes inadequate. The concentration of these nutrients in old leaves gives a good indication of nutrient status, whereas the mobile nutrient level of the developing young leaves is a poor indicator.

Immobile nutrients, such as Ca, are incorporated in plants only from the external medium. Deficiency of Ca may occur even if an excess of Ca is found in older tissues of the plant, when the supply of Ca in the nutrient solution has been reduced. The amount of Ca in young leaves may be used as an indicator of Ca supply to the plant.

Boron is also considered immobile from older plant tissues to developing leaves.

An intermediate status of movement of nutrients in the plants occurs for Fe, Zn, Cu, and Mn. Some of these ions are mobile (through the phloem system); the degree of mobility varies with plant species, environmental conditions, and stage of plant growth.

When the concentration of an ion in the plant is high, the mobility of this ion is higher than when the concentration is near the deficiency level.

The relations between ions may also interfere with the mobility and the supply. A deficiency of one ion may affect the efficiency of a related ion even if it is present in relatively fair amounts, and symptoms of deficiency of this more abundantly supplied ion occur. Such a relationship was observed between sulfur and nitrogen.

3.12 Nutrient Concentration and Yield

It is generally believed that balanced nutrient solutions, even those deviating from the recommended concentration, have very little influence on growth or yield. A much more pronounced effect is produced by the deficiency or excess of any single nutrient than the overall effect of the total nutrient solution. Similarly, the deficiency or excess of a specific ion has a pronounced effect on the yield or total dry matter production. Figure 3.2 illustrates the optimal and surplus ranges of ion concentration as related to yield or production of dry matter. Boron differs from other elements with a narrow optimum range.

3.13 Preparation of Nutrient Concentrates

In modern soilless culture systems, where the nutrients are supplied via injection pumps or proportionating dilutors, it is usually necessary to prepare concentrated stock solutions of fertilizers at 50 or 100% the final concentration. Where automated nutritional control is based on conductivity measurements, the overall concentration of the stock solutions is unimportant, provided that the ratios between nutrients (and hence the relative concentrations of solutions A and B, see below) remain unchanged.

If all the fertilizers required in the final solution were mixed together in concentrated form, precipitation of sparingly soluble salts such as calcium phosphate or sulphate would result. To avoid this two separate

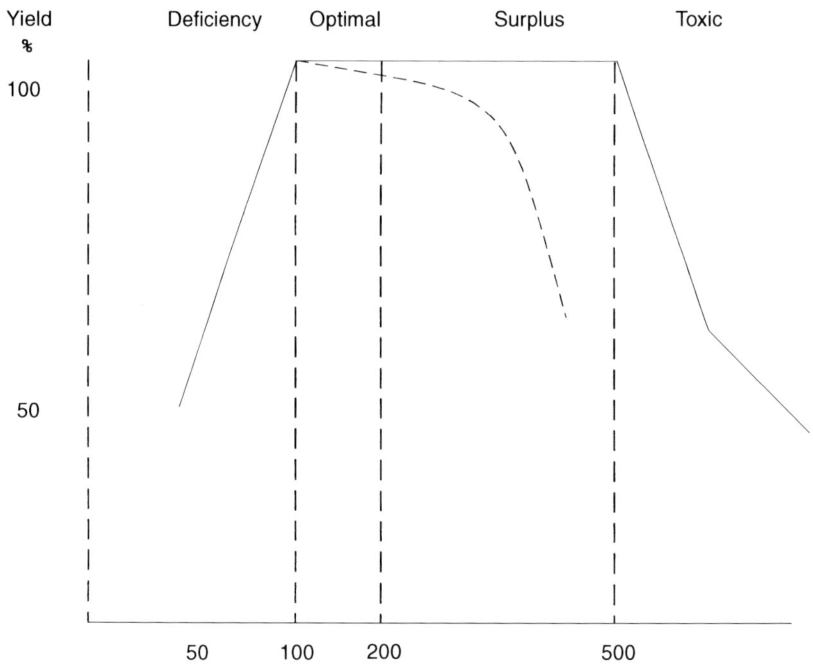

Fig. 3.2. Yield or dry matter production according to the amount of a specific ion in the nutrient solution. Concentration of ion in nutrient solution in % (100% = the lowest level of ion to receive 100% yield. ——— Any nutrient ion; - - - - boron ion

stock solutions are prepared, often with a third solution containing acid for automated pH control.

Early formulations placed the calcium nitrate in solution A and the remainder of the chemicals (including phosphates and sulphates) in solution B. More advanced formulations are now in use, however, in which part of the total acid requirement is provided as nitric acid in solution A, together with the chelated iron. The total requirement of potassium nitrate may also be split between solutions A and B. The remaining fertilizers, including any phosphates or phosphoric acid, magnesium sulphate, potassium sulphate and micronutrients other than iron, are grouped in solution B.

Additional nutrient solution preparation advice is given in Chap. 4 on the various culture methods.

3.14 Foliar Nutrition

Certain ions, such as nitrogen, iron, and zinc, can be supplied through the leaves. By applying these ions provided in fertilizers to the leaves, quick recovery from a nutrient deficiency has been obtained.

If nutrients can be absorbed through leaves, some should also be lost from leaves by leaching due to rainfall, and indeed some mineral salts are lost in this way. Potassium appears to be particularly easily lost by leaching, perhaps because of its high mobility in the plant (Alexander 1986).

3.15 Nutrient Solution Analysis

In order to realize the full benefit of soilless culture, the nutrient solutions should be tested frequently.

In open systems (such as sand culture or rockwool culture), the samples should be taken from the dripper or the end of the irrigation pipes before the solution reaches the plant. Samples should also be taken from the drain-waste nutrient solution.

Errors in the preparation of the nutrient solution are not uncommon, making an analysis necessary to check the final elemental concentrations prior to use. Since the elemental composition of the nutrient solution can be altered considerably in open, recirculating systems, it is equally important to monitor the composition of the solution as frequently as practical. A record of the analysis results should be kept. On the basis of such analyses, change schedules, replenishment needs, and crop utilization patterns can be determined.

In closed systems, such as NFT or water culture, the nutrient solution testing is even more important.

Water minerals are not taken up from a nutrient solution in the same proportion in which the components are present. Rate of absorption depends upon the plant, climatic and environmental conditions such as light intensity and duration, temperature, and humidity, as well as the type of culture and its stage of development. Uptake of water and nutrients by the plants will constantly change the nutrient solution, increasing the concentration of salts disrupting the balance of the components. Hydrogen ion concentration (pH) will also change as a result of osmotic pressure, differences in uptake, and possible reaction with the aggregate and root secretion.

Addition of tapwater or rainwater to the nutrient solution to replenish loss by transpiration will bring the nutrient solution to its original volume. Testing of this diluted solution will indicate which elements should be added to this water if at all. The testing also enables a longer and thus more economic use of water and nutrients. It is often important to test for accumulation of undesired elements or required elements developing in toxic concentrations, such as sodium.

Type, procedure, and frequency of tests will depend on farm size, the crops grown, special conditions such as salinity, and equipment available. If the amount of solution is small and chemicals relatively inexpensive, it may be easier, cheaper, and quicker to change the entire nutrient solution instead of testing to determine the cause of poor growth.

Some remarkable developments in recent years have made routine analysis of the nutrient solution easier and cheaper. Laboratory analysis is recommended, although on-site analysis is possible with the use of kits and simple analytical devices. It is now also possible to continuously monitor the nutrient solution with devices such as specific ion electrodes. Therefore, irrespective of the growing system in use, some type of analytical procedure should be used to monitor the composition of the nutrient solution as a normal routine.

The analysis of the nutrient should include pH, and tests to determine the concentration of the major elements, nitrate (NO_3^-)- and ammonium ($NH4^+$)-nitrogen (N), phosphates (P), potassium (K), calcium (Ca), and magnesium (Mg). All of these determinations can be easily made on site using a water analysis kit.

Although similar kits are also available for some of the micronutrients, laboratory analysis will be necessary in most instances. However, concentration monitoring of the micronutrients is not as critical as for the major elements unless a micronutrient problem is suspected. For any diagnostic problem, laboratory analysis is always recommended, including all the essential elements – the major elements and micronutrients.

The most frequent routine tests for nutrient solutions are pH and total salt concentration (chap. 9). Less frequent tests are nitrogen [such as nitrate (NO_3^-) and ammonium (NH_4^+)], phosphorus [if possible also potassium (K) and calcium (Ca)], and dissolved oxygen (DO).

The quickest and most convenient applied tests are done using commercial test kits. These are available as separate or combined units. The procedures are simple and the equipment inexpensive.

4 Culture Methods

4.1 General Outlines

As previously discussed, soilless culture is the term most frequently used to include all nonconventional soil culture methods, although the term hydroponics is often still found in the literature. Both terms can be used to describe all the general nonsoil growing systems.

In some countries the term hydroponics is still coined for the original water culture method. As water culture methods were not found economical and many investigators lost money, the name hydroponics developed a negative connotation and investigators tended to avoid using the term. In other countries the term hydroponics refers to advanced agriculture and connotes progress and novelty to the new agriculturists, while soilless culture is meaningless to them.

Laymen often have prejudices against soilless culture. To them the term hydroponics means "it requires a lot of water" or "it requires a small amount of water," "it is a waste of money," "the vegetables taste bad," etc. As these VIPs are often the decision-makers, it is important to choose the right "understandable" term when discussing agriculture. Failures of soilless-culture projects have often resulted simply from poor choice of descriptive terms.

The managing of current soilless-culture methods is the topic of this book. Therefore this chapter, rather than looking at the historical development of water culture or discussions of the various controlled methods (from the more controlled methods such as water culture and closed systems to the less sophisticated methods such as organic mixes and open systems), will be devoted to detailed descriptions of the current five most convenient and economical methods. Development of the various methods is dynamic, and what today seems best may be rejected in favor of new advanced systems tomorrow.

The five main soilless-culture methods used commercially are the:

1. Sand culture method – open system (Jensen 1980, 1981, 1984; FAO 1990)
2. Rockwool method – open system (Van Os 1982)
3. Gravel culture method – closed system (Schwarz 1968; FAO 1990)

4. NFT and water culture method – closed system (Graves 1983; FAO 1990)
5. Bag culture methods – open systems (Judd 1982; Maree 1981a)

4.2 The Supporting Media

4.2.1 Inorganic Media

4.2.1.1 Sand

Sand is usually stated as particles below 3 mm in diameter (Steiner, 1976). Pure sand can be used in trough and trench culture. However, in desert locations, it is often more convenient (and less expensive) to cover the greenhouse floor with polythylene film and a system of perforated drainage pipes, and then backfill the area with sand to a depth of at least 30–35 cm. If the depth of the sand bed is shallower, moisture conditions may not be uniform and plant roots may grow into the drain pipes. The area to be used as planting beds may be level or slightly sloped; supply manifolds for nutrient solution must be sited accordingly. The sand culture method is described in Sect. 4.3.

Different types of desert and coastal sands with a variety of distribution of sand particles are not critical, except that exceptionally fine material, such as mortar sand, does not drain well and should be avoided.

The finer the sand particles, the deeper (up to 40 cm) the layer of sand required. Sand grains will constantly hold water at the lower part of the layer from the bed bottom up. This phenomenon is similar to any regular sponge filled with water to full capacity. While the sponge is held up, the lower part of the sponge – several centimeters – is wetter than the upper part. Where the sponge is attached to a surface, part of this water will drain, but will still behave, in regard to water capacity, similarly to a sponge. This lower sand layer is a poor, unaerated medium for root-system development.

Calcareous sand can be used with a specially adapted nutrient solution. This is possible for certain crops. See section on nutrient solution composition (3.8).

Sand growing beds should be fumigated annually because of possible introduction of soil-borne diseases and nematodes.

4.2.1.2 Gravelly Material

Basalt or granite gravel, or riverbed flint aggregates, are good supporting media. Porous material such as lava is even better, due to its greater water-retaining capacity; this may be particularly important in tropical areas where the transpiration rate is high.

Particles should be about 0.75 cm in diameter, and free of fines, since these would accumulate on bed floors. Nonuniform aggregates should be sieved to required size. Where separate beds are used for seeding nursery plants, the aggregates should be about 0.5 cm in diameter. Soft aggregates should not be used.

Beds should be filled to a depth of 18–20 cm for most crops, but to 25 cm for deep-rooting crops such as roses, grapes, etc.

Tuff gravel, or scoria (Fig. 4.1), is one of the best types of gravel. Even particle sizes of 1.00–1.25 cm can be used. The color of the tuff gravel, usually a result of the degree of oxidation of the gravel, has no influence on growth.

Scoria is not inert. It has certain cation and anion absorption properties having no deleterious effect on plant growth. The good water-holding capacity of scoria and the air exchange possibilities provide an excellent plant growth material.

Calcareous (marble or limestone) gravel should be avoided. If this is the only available material, it may be used together with a special solution.

Research showed that a constant nutrient solution level of 1 mM of phosphorus (by daily addition of 0.2 mM PO_4), instead of 3 mM, enabled

Fig. 4.1. Tuff (scoria) culture in Israel

nonpretreated calcareous gravel to give results almost equal to those for granite gravel (Schwarz and Vaadia 1968).

A higher iron level (usually double the amount) is required when using calcareous gravel.

Black and dark gravel absorb more heat than white gravel and may harm young plants at the root crown section. Older plants shade themselves, and the color of the gravel does not affect them.

4.2.1.3 Vermiculite

Vermiculite is a hydrated magnesium aluminum silicate which occurs in plate-like crystals.

When heated, the material expands greatly at right angles to the plane of cleavage. This process is known as exfoliation (Bently 1959). Portable exfoliators have been developed for use in countries far from the primary production areas, since storage and delivery of the low-density material can be costly. The treated (exfoliated) product is a light-weight material having an average bulk density of about 80 kg m^{-3}. The material sometimes has an alkaline reaction due to the presence of magnesian limestone in the original ore. Both the exchange capacity and the water-holding capacity of exfoliated vermiculite are high. The material thus has many of the characteristics of a good hydroponic substrate.

A reduction of 20% density of the vermiculite structure is expected after a 1 year of use. After 2 years, it can no longer be used as it will collapse and lead to reduced aeration and drainage.

The high cost makes this substrate not commercially viable. It is used in plant nutrition research and as a part of growth mixture media.

4.2.1.4 Foam Plastic

Polystyrene foam plastic is the most useful; since it is inert, it has some water-retaining capacity, and is extremely light.

It is advisable to water foam plastic for 2–3 weeks before planting, because formaldehyde may be given off and this may reduce yields.

About 3–5% of foam plastic remains attached to the roots with each drop removed, and is thus lost.

Other types of foam plastics used in soilless culture, especially in bag culture methods and as mixtures, are polyurethane, urea-formaldehyde, and polysterene.

As most of these are lighter than water, they can be used in open beds. The mixture in bags is usually combined with heavier material, such as sand, peat or wood-bark.

During the growth period, the ingredients of the original mixture change position: the heavier material, such as sand, settles to the bottom, and lighter materials rise to the top.

Table 4.1. The properties of foam plastic urethane and phenolic used as growth medium. (Boodley 1984)

	Urethane	Phenolic
Density (hg/m^3)	15	28
% moisture		
weight	464	376
volume	93	98
drainage	55	32
pH	6.9	6–7.5
Ec (mmhos/cm)	<0.2	<1.4
color	Black	Green

In the USA, several foam plastic materials are available. The properties of two well-known materials are given in Table 4.1.

Polyurethane slabs are an excellent growth media and it is used similar to rockwool slabs. It was found stable for more than six years (in Belgie).

4.2.1.5 Perlite

Perlite is derived from a siliceous volcanic rock. The material contains some 2–5% moisture and, when crushed and heated to about 1000 °C, expands to form a lightweight material of bulk density 130–180 kg m^{-3}. The material has a closed cellular structure, most of the water retained after irrigation being held superficially and released at relatively low moisture tensions.

Substrates containing perlite as a major component are thus well drained and well aerated. The product is physically stable and chemically inert. It contains 6.9% aluminum (Al), however, and some of this may be released into solution at low pH, with unfavorable consequences for plant growth (Green 1968; Wilson et al. 1984).

Perlite is now widely used as a component of peat-based potting substrates (Wilson et al. 1984; Hall et al. 1988).

4.2.1.6 "Kuntan"

Composed of carbonized rice hulls, "Kuntan" is prepared by baking the rice hull raw material. In areas like Japan, where large quantities are available, this medium is used for soilless culture in beds or bag culture (Namioka 1976).

4.2.2 Organic and Mixed Growth Media

Soilless culture is deeply interwoven with anorganic growth media and anorganic nutrition. The use of anorganic growth material provides a precise control over most of the growth factors. Some factors are more favorably affected by the use of organic material. Examples of such factors are pH-buffering effects and water- and air-holding capacities. Disease and infection, however, are more prevalent when organic materials are used.

Pure organic materials, such as peat or manure, although not technically soil, are not considered to be soilless-culture growth media.

Mixtures of organic and nonorganic growth media, however, are included in soilless-culture growth media descriptions and therefore deserve discussion. The bag-culture systems often employ soilless-culture mixes which include organic material. These mixtures are usually irrigated with a nutrient solution as is done when anorganic growth media are used (Maas and Adamson 1981).

Some growers prefer the addition of fertilizers to the growth media mix, or the use of slow-release fertilizers, methods which are not recommended because of lack of control over the balance of nutrients.

Growth of plants in containers and pots (bag culture) with single or few plants in each container has advantages: it allows for the removal of unhealthy plants (containers) and forestalls infecting the entire system; each individual container is served by its own separate drip system, unassociated with the neighboring plants and their water-nutrient balance control. This necessitates a large reserve of water, which is achieved by adding organic water-holding material to the mixed growth culture.

Organic material is usually defined as that which serves as a substrate for biological processes and is subject to biological decomposition. Although polyethylene, polyurethane foams, and polystyrene are composed of organic materials, they are not subject to biological decomposition and are therefore regarded as anorganic.

Organic material for use in soilless culture can be divided into three groups:

4.2.2.1 Peat

Peat is formed as a result of partial decomposition of plants under anaerobic or semianaerobic conditions. Peat has a low pH and low levels of nutritional elements.

Peat bogs are formed under climatic conditions of high precipitation, low evaporation, low solar radiation, and low temperatures.

There are several types of peat currently used. Sphagnum peat has a high water-holding capacity. Sedge peat, with a lower water-holding capacity but a higher air-holding capacity, is more stable under aerobic (greenhouse) conditions and decomposes more slowly than sphagnum peat. Both types of peat have low humic acid content (about 5%).

Hypnum peat moss (formed as a result of Bryales or woody plant degradation) contains high concentrations of humic acids. The humic acid can be precipitated by a liming process which forms nonsoluble humate salts. Part of this lime can be released while irrigating with a nutrient solution of low pH.

The high cost of peat, especially in countries lacking local peat resources, decreases the cost efficiency of using this growth medium or mixtures containing it for commercial plantations (Adams et al. 1989).

4.2.2.2 Peat Substitutes

Bark. Bark protects the inner part of trees against desiccation and phytopathogenic organisms. Through the use of debarking machines, bark is stripped off logs in the sawmill and then ground by hammer mills and screened.

Softwood bark (from coniferous trees) comes mainly from *Pinus* spp., *Picea* spp. (spruce), *Cypressus*, and *Larix* spp. (*Cedrus*).

Hardwood bark trees include *Quercus* spp. (oak), *Fagus* spp. (beech) and *Betula* spp. (birch).

Bark is processed by grinding, composting, and sieving. It has a low water-holding capacity and a high air-holding capacity and is therefore similar to regular coarse soilless culture growth media. The degradation rate of bark is slow.

Some barks may release manganese which, in severe cases, can produce manganese toxicity symptoms.

Sawdust. Sawdust and woodchips come from the inner part of the tree. The carbon to nitrogen ratio is higher than that of bark; the lignin content is lower; and a longer composting period is required.

The use of wood waste as part of the growth media without previous composting is not advisable. The composition of wood waste is for the most part not uniform, and occasional phytotoxicity problems have been reported. It should be used for not more than 15–20% of the total media mix (Adamson and Maas 1971).

Porosity and Aeration. The recommended total porosity in most mixed substances is above 80% by volume. The substrate pore spaces contain water and air. The air content is the proportion of the volume that contains air after it has been saturated with water and allowed to drain. As growth usually takes place in containers or bags, often the lowest part of the container still holds water that has not fully drained in spite of the drainage holes underneath. The height of the container/bag affects the drainage capacity also. It is therefore important to measure and know the porosity and aeration capacity for each specific growth case. This knowledge is especially important in warm climates when growing plants with high water requirements or salt sensitivity. The growth media volume and the pore-media volume should provide enough water-holding, air-holding and water-air exchange capacity.

Sewage Sludge, Municipal Compost, Animal Excreta. These are not recommended mainly because of the considerable heterogeneity of the products, phytotoxicity problems, high salinity levels, and low air-holding capacity.

Sand. Coarse sand can be added to up to 20% of the volume. After prolonged irrigation, the sand will settle in the lower part of the mix.

The recommended water pH for an organic mix is 5.0–5.5. By adjusting the pH of the nutrient solution irrigation the required pH in the root environment can be controlled. Addition of fertilizers to the organic mix is not recommended.

4.3 Sand Culture – Open System

Two methods of utilizing sand as a growing medium are usually recommended. One is the use of polylined or similar troughs, the other involves spreading sand over the entire floor (Johnson 1980; Jensen 1984; Schwarz 1968, 1984).

4.3.1 Troughs Lined with Polyethylene

Growing beds may be built as aboveground troughs with wood, concrete, timber, fiberglass on asbestos sides. The entire trough is lined with a layer of approximately 0.15–0.20 mm-thick polyethylene to separate the growing medium from the soil below. The bottom of the trough should have a slight grade of 15 cm per 30 m, so that it can be drained or leached when necessary. The drain pipe, 1.25 in. or more in diameter, is laid from one end to the other on the bottom of the trough. Asphalt-coated drain pipe, PVC, or any type of drainage pipe is used. Drainage holes are cross-cut, one-third of the distance through the pipe every 50 cm. If holes are drilled in the pipe, they should be covered with fiberglass or similar material. The pipe should be at least 4 cm in diameter, and the holes should be located on the underside, i.e., near the bottom of the bed.

The depth of the beds varies with the type of substrate. Using sand with particles in the range of 0.3–0.5 mm diameter, a depth of at least 35 cm is required. As already mentioned, the need for adequate depth of bed arises because the lower part of the substrate will be saturated with water and is thus a poor medium for root development. The use of deeper beds helps to ensure that at least the upper part of the substrate will not be waterlogged after irrigation. The bottom of the bed, from side to side, may be either level or V-shaped with the drainage pipe in the middle in either case. The beds are then filled with coarse sand.

4.3.2 Whole Greenhouse Sand Culture

The second possibility is to line the floor with polyethylene and fill with sand (Fig. 4.2). The floor must have a slight grade of 15 cm per 30 m, so that the growing bed can be drained or leached when necessary. A 0.2 mm polyethylene plastic or two 0.1 mm sheets is used to cover the entire floor.

Prior to installing the polyethylene, the floor would be graded and packed with a roller and any low spots filled and compacted to the desired level. When a floor is wider than the sheet of polyethylene, the sheets may be overlapped approximately 30 cm.

Drain pipes are then placed on the floor so that there is a uniform spacing of approximately 1 m between pipes. If the drainage rate of the sand is less than that for normal sand, the pipes should be closed. In either case, the drain lines should lead down the slope into a main drain running across the low end of the grade. Any drainage can then be directed to a sump and pumped onto trees, etc.

Drainage holes are cut into the pipe. Once the plastic and drain pipes are in place, sand is spread over the entire area to a depth of at least > 35 cm, as already mentioned.

A regular drip irrigation system is used. Each time water is added to the sand, a complete adequate nutrient solution is always added. The fertilizers may be fed into the irrigation water through proportioners or mixed into the water in a large tank and then pumped directly onto the crop. All watering may be programmed through the use of a time clock. In a large installation time clocks may be connected to solenoid valves which allow only one section to be watered at a time.

This technique was developed by Jensen (1981, 1984) in Arizona, and is used in Israel, the USA, Peru, and recently in many new installations.

4.3.3 Sand-Sprinkling Systems

The ultimate extension of low cost culture is the use of sand dunes for agricultural purposes. The technique, described by Schwarz (1968), is basically similar to the "whole house sand culture" described in the preceding paragraphs. The beds are prepared with a minimum depth of 35 cm, with improved root environment if depths of 60 to 70 cm are possible. The floor is graded and compacted to give a uniform slope of 0.5%, and covered with polythene sheet (0.15 mm). If the area is wider than can be covered with a single sheet, the material is overlapped for about 30 cm. Drainage pipes are laid on the floor at a spacing of 1 m, or closer if the sand does not drain readily; the drainage is collected in a transverse main drain at the bottom of the slope. Nutrient solutions are applied via a drip-irrigation system. Watering can be timer-controlled,

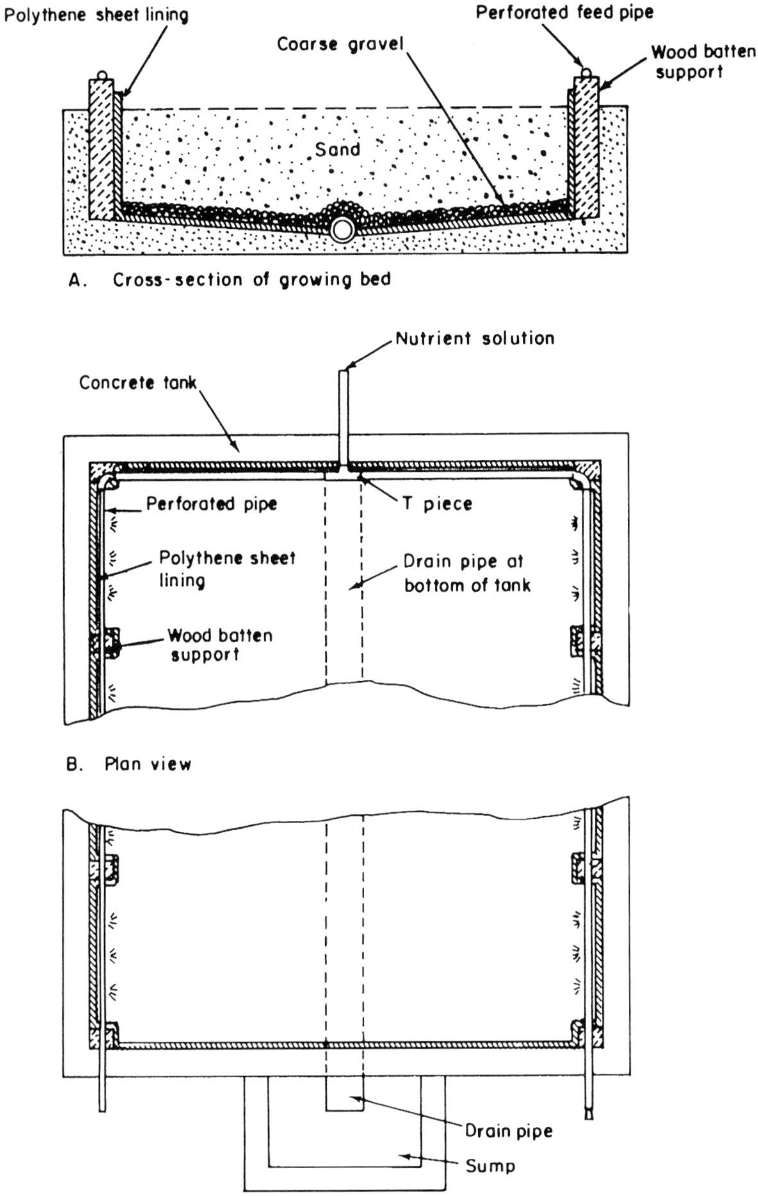

Fig. 4.2. An alternative open circuit system in sand with overhead irrigation (Stoughton 1969). Drainage is towards a sump at the lower end of the bed

and solenoid valves can be used if required to water one section at a time on large installations.

4.3.4 Sand Dunes

Where the sand dunes are more than 1 m deep, construction of beds is unnecessary, since the root zone will not normally be too wet. This system has recently been used in Israel.

Sand beds require regular, usually annual, sterilization to eliminate soil-borne diseases and pests (including nematodes). Chemicals listed for such purposes by Schwarz (1968) included dilute solutions of formaldehyde (1%) and calcium or sodium hypochlorite (0.3–1.0%); the solutions are allowed to stand in the beds for 24 h, after which the beds are leached three to four times. The surface of the beds is completely covered with polythene sheet before treatment is started. Great caution is needed when using any such chemicals because of the risk both to workers and to other crops in the vicinity. Methyl bromide, a gaseous fumigant, is particularly effective but highly dangerous; its use is governed by regulations in most countries, and professional advice must be obtained and all necessary precautions taken.

The watering system, must be flushed with water after application. The beds be replanted four-five days after fumigation, By using steam sterilization, viruses are killed as well. Steam may be put through the drainage system.

Watering Schedule. The crop may be watered one to three times per day, depending on the age of the plants and time of the year. Enough water is added to allow a small trickle of solution to drain off the entire growing area. A sample of this drainage can be collected and tested for total dissolved salts. If this drainage water has a high salt reading (possibly due to extraneous salts such as sodium), the entire bed is leached free of salts, or to a point nearly equal to the salt level of the irrigation water. If a fertilizer proportioner is used, the total dissolved salts of the nutrient solution going onto the plants should be checked each week to see whether the injector pumps are working correctly. Also check on occasion as to whether each injector pump is injecting the proper amount of stock solution into the irrigation water. It might be that one pump is injecting too much, while the other is injecting too little.

Fertility levels maintained in the sand depend greatly on those levels in the nutrient solution. The nutrient status of a crop may be checked by plant tissue analysis.

Sand culture is not new, but now that plastics are common in agriculture, along with commercial water-soluble fertilizers, it appears to be both practical and economical. It is a system which provides excellent control over diseases and allows maintenance of optimum water and fertility conditions with precision and with very little effort.

4.3.5 Alternative Growth Media

It has recently become popular to use this simple growth system with locally available inert growth media such as gravel, tuff, etc. When using larger particles (0.5–1.0 mm), the depth of the beds can be reduced to 20 cm as in gravel culture methods.

Compared to ordinary sand culture, more frequent irrigation is required according to the type gravel, plant, season, etc.

4.3.6 Differences Between Sand and Gravel

The most important difference between sand and gravel culture is that in gravel culture it is possible to circulate the nutrient solution in a horizontal direction, while in sand culture the pores are too small for such a rapid transport. This difference is fundamental when a subirrigation procedure is used and nutrient solution is reused, (see gravel culture Sect. 4.5). When trickling or triple irrigation is practised as in the sand culture method, both substrates can be used and duration of watering is adjusted accordingly.

Another difference is that after irrigation in gravel culture about 15% of the intergravel space is filled with air, while in sand culture it will be less.

However, adjusting the irrigation to avoid sand that is too dry, resulting in wilting of the plants, or too wet, resulting in lack of oxygen, is one of the main problems in sand culture compared to gravel culture.

4.3.7 Trough Culture with Organic Substrates

The type of bed described in Sects. 4.3 and 4.3.1 is also sometimes used for crop production in substrates consisting partly or entirely of organic materials such as peat, sawdust or wood bark. Mixtures of peat moss and vermiculite with added nutrients (based on Collins and Jensen 1983) have proved satisfactory for trough culture in the USA.

Troughs filled with peat have been used extensively in Finnland and other Scandinavian countries, and to a lesser extent in the Channel Islands and Europe including the UK, though in the latter country cropping in peat bags (see bag culture) is more widely adopted. A description of commercial tomato growing in troughs of peat has been given by Potter (1977). The cheapest form of trough is made up from 500 gauge black polythene, turned up at the sides and supported by wires to form a bed 0.76 m wide and 10 cm high. Drainage is not always provided, but the floor should be carefully levelled and the row length should preferably not exceed 18–20 m. Watering must be controlled with care, and is often applied with a low-level sprayline for even coverage.

As open systems are less sensitive than closed systems to the type of growing medium used, a great deal of regional ingenuity has been displayed in locating low-cost inert materials for trough culture. Typical media include sand (Jensen 1984), vermiculite (Harris 1976), sawdust (Adamson and Maas 1981; Maas and Adamson 1981), perlite, peat moss (Maher 1976), mixtures of peat and vermiculite (Jensen 1968), and sand with peat or vermiculite (Smith 1982). One of the most effective – but expensive – of these materials is a mixture of peat moss and vermiculite developed at Cornell University (Sheldrake and Boodley 1965) to provide trough growers with a sterile soil. The peat moss-vermiculite medium is customarily used with drip or trickle irrigation systems; a fertilizer regime for plants grown in this medium is discussed in Sect. 4.8.17.

4.4 Rockwool Culture – Open System

4.4.1 Description

Rockwell or mineral wool is a man-made inorganic material which is used as an inert substrate for crop production. Forms of mineral wool find extensive application as lightweight materials for acoustic and thermal insulation. The horticultural use of slightly modified types of mineral wool was developed in Denmark in 1969.

Rockwool culture usage has been reduced recently in Europe, particularly in the Netherlands due to environmental problems. The greenhouse area devoted to rockwool culture in the Netherlands is about 3000 ha, in Canada 100 ha, and in Israel 160 ha. Rockwool is primarily used in "open" (nonrecirculating) systems of hydroponics.

Although rockwool is also available in "granulated" (fibrous) form as a component of potting mixtures (Verwer and Welleman, 1980), the term rockwool culture is mainly applied to crop production in slabs and blocks of the bonded material. The blocks are firm enough to be handled yet sufficiently permeable to allow root penetration.

4.4.1.1 Manufacture

The material is manufactured by heating a mixture of 60% of the igneous rock known as diabase (dolerite) with 20% each of limestone and coke. The coke acts as a fuel in a form of blast furnace (cupola), through which air is forced to raise the temperature to about 1600 °C. The rock melts and is tapped off from the base of the furnace on to a series of high speed rotors. The molten droplets which spin off the rotors form fine fibers which are then cooled by an air stream. During the later production stages a phenolic resin is added as a binder, and a wetting agent is also incorporated. The product is pressed, allowed to harden, and cut to shape.

The chemical composition of the material, expressed as the oxides, is given (Verwer 1976) as:

SiO_2	47%
Al_2O_3	14%
CaO	16%
MgO	10%
Fe_2O_3	8%
Na_2O	2%
K_2O	1%
MnO	1%
TiO_2	1%

Rockwool as generally available has a bulk density of about 70–100 kg/m^3 and a pore volume of about 96%.

4.4.1.2 Water Relationships in Rockwool

In using mineral wool for growing plants it is important to recognize that the material has a high water retention at low water tension, and that the water content will increase from top to bottom in the block. For example, if the material is saturated and allowed to drain, the air volume will be only 4% at a height of 1 cm above the base (Verwer and Welleman 1980). This can cause problems of rooting if the height of the water column is less than about 5 cm, as can be the case during propagation in small blocks. To overcome this problem during the rooting of cuttings Verwer and Welleman (1980) recommended that the substrate should be moistened but not saturated, i.e., that the water applied should be only about half the volume of the rockwool, the small blocks should be placed on a drainage layer of sand or perlite and the cuttings should not be inserted deeply into the material. The latter recommendations would apply equally to the germination of seeds if the latter are inserted deeply into holes made in the top of the block.

When the young plants (e.g., tomatoes, cucumbers) in their propagating blocks are subsequently placed in position on the rockwool slabs, water tension at the top is increased (i.e., water content is reduced) and good aeration is ensured.

Plants are usually propagated by direct seeding in small (40 mm^3) rockwool cubes that have a small hole punched in the top and are saturated with nutrient solution. These cubes are usually transplanted into larger rockwool cubes (75 mm^3), manufactured specifically to receive the germinating cubes, and side-wrapped with black plastic film. The larger cubes are subsequently placed on top of rockwool slabs placed on the greenhouse floor. The slabs are usually 15–30 cm wide, 75–100 cm long, and 75 mm thick.

In a typical rockwool culture, the level ground is covered with white polyethylene film for good hygiene and light reflection. A bed normally consists of two rows of rockwool slabs, each individually wrapped in white film, with the rows spaced 30 cm apart. The slabs should have a slight inward tilt toward a central drainage swale. Because of the porosity of the rockwool, almost all of the nutrient solution remains in the slab for plant use if an appropriately modest irrigation schedule is used; if there is a surplus, it will drain out of the slab and into the shallow channel.

Before seedlings are transplanted, the rockwool slabs are soaked with nutrient solution. The plants, remaining in the small rockwool cubes in which they were established, are simply set on top of the slabs. If the root system is well developed in the cubes, the roots will move into the slab within 2–3 days. Each plant receives nutrient solution via individual drippers, with irrigation rates varied as a function of plant demand and environmental conditions.

4.4.1.3 Sizes, Wrapping, and Layout of Rockwool Slabs

When rockwool culture was first introduced the slabs used for row crops were 100 cm long, 30 cm wide, and 7.5 cm deep. The slabs were laid out in rows on polythene sheet of sufficient width to wrap up round

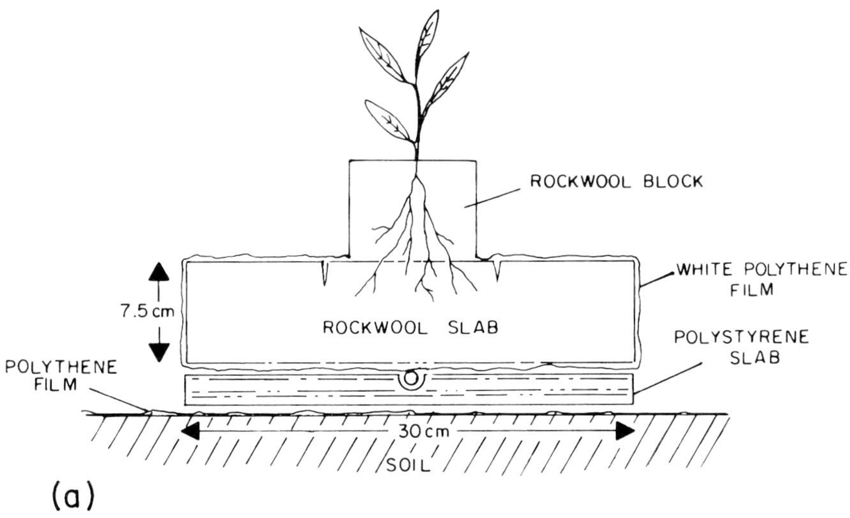

Fig. 4.3. Generalized layouts for crop production in rockwool. **a** Single row crop. **b** Double rows with drainage channel (FAO 1990)

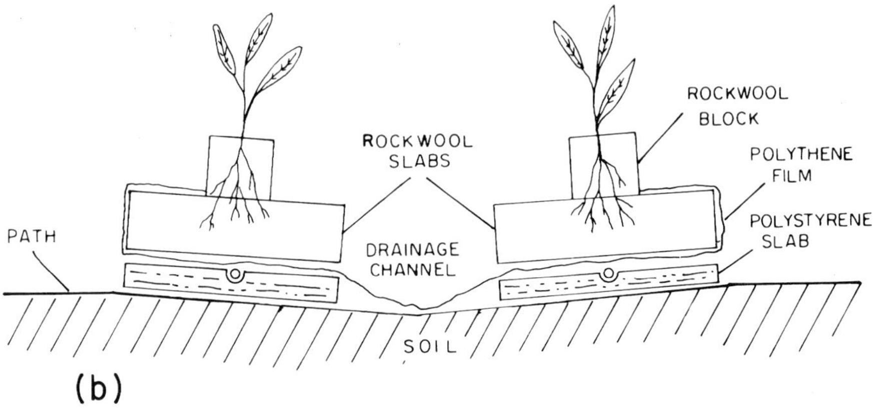

Fig. 4.3. *Continued*

them, covering the tops except along the center where the propagation blocks would be placed. The general arrangement, with heating pipe and underlying polystyrene sheet, is shown diagrammatically in Fig. 4.3. A possible layout for tomatoes, grown in twin rows is shown in Fig. 4.4, based on Hanger (1979); a single sheet of polythene partially encloses both slabs and provides a drainage channel between them.

The 30 cm-wide slabs were at first often considered economic for cucumbers but hardly so for tomatoes. The latter crop has a less extensive root system, however, and it was later realized that narrower (15–20 cm) slabs would suffice, thus greatly improving the economic assessment. A layout for tomatoes or roses that is now widely used commercially is shown diagrammatically in Fig. 4.4 consisting of the narrower (15 cm) slabs, fully prewrapped and laid on a flat surface below the level of the paths. The drainage slits are made on the sides of the bags away from the

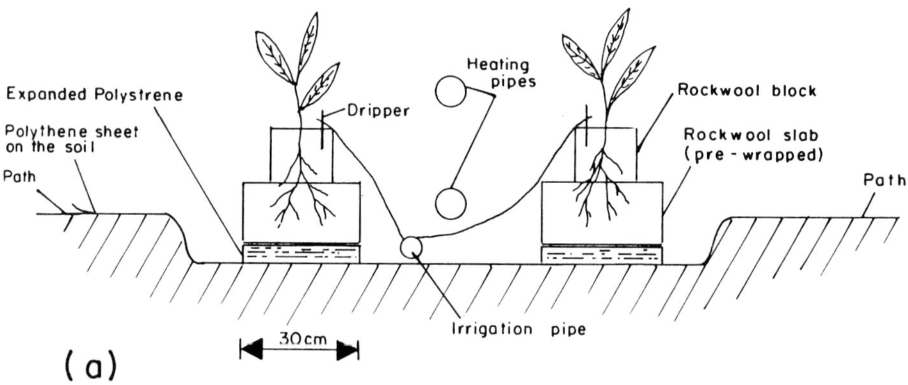

Fig. 4.4. Rockwool culture (diagrammatic). **a** Modern commercial tomato production. **b** Rose growing (FAO 1990)

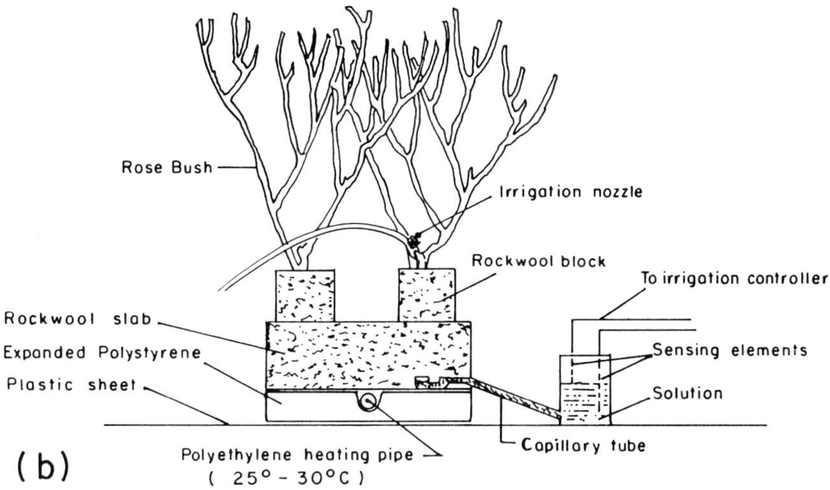

(b)

Fig. 4.4. *Continued*

path. No drainage channel as such is provided; holes are merely punched in the polythene floor covering at any point at which surplus solution is seen to accumulate.

The rockwool may be fully wrapped in white polythene film (Fig. 4.5). Figure 4.5 also shows an interesting device for controlling the irrigation system, known as the Niewkoop pot. This control device is connected to the rockwool slab by means of a short length of capillary tubing which is inserted through the polythene wrapping. The tube contains a length of very thin glass-fiber material, rolled up to fit inside it; this restricts the rate at which water moves in and out of the pot and thus stabilizes the control system. The control device contains two electrodes, the lower of which is always in contact with the nutrient solution. The equipment is so adjusted that, when the solution level falls below the upper electrode, the irrigation system is switched on.

Commercial production of greenhouse roses in rockwool is illustrated in Fig. 4.6 the rockwool slabs seen here are supported on a metal grid above ground level.

Rockwool is available in the form of slabs, blocks, or granules. The slabs are rectangular pieces of varying length and width but are usually 7.5 cm high for raising long-season crops such as tomatoes, cucumbers, peppers, eggplant, etc. Typical dimensions are 90 × 15 × 7.5, 90 × 20 × 7.5, and 90 × 30 × 7.5 cm or longer. The blocks also come in a variety of sizes and are used for seed germination and transplant raising. The granular form is used for soil or soilless-mix amendment, and can also be used in bag culture as a partial or complete substitute for peat (Donnan 1988).

Fig. 4.5. Roses growing in wrapped rockwool slabs supported on a metal framework (FAO 1990)

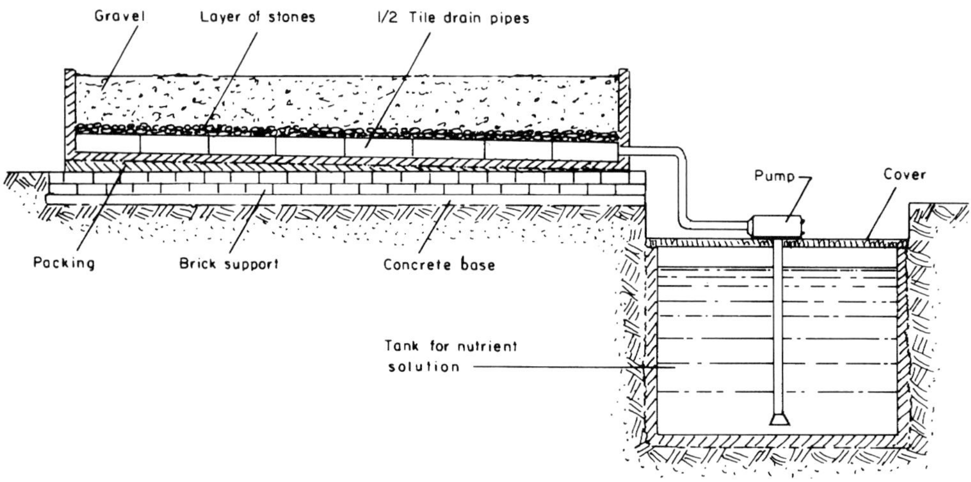

Fig. 4.6. Subirrigated gravel culture (direct feed) (based on Stoughton 1969). Alternatively, the solution is pumped to the far end of the bed and flows back down the slope to part-fill the bed, returning to the tank when a valve in the drainage system is opened (FAO 1990)

4.4.1.4 Propagation and Early Growth

Plants intended for growing on rockwool slabs should be propagated in an inert medium in order to take full advantage of the initial sterility of the material. The seed may, for example, be sown in vermiculite or perlite and pricked out into the blocks (e.g., 0.42 l) as soon as the seedlings can be handled. Alternatively, a small rockwool cube can be used for pricking out or even for direct germination, this in turn being set in the hole of a larger (0.42 l or larger) block. The blocks are thoroughly wetted with the appropriate nutrient solution before use, and are stood on coarse perlite above polythene sheet to ensure good drainage. The blocks are finally set out on top of the rockwool slabs, again thoroughly wetted with nutrient solution. A drip irrigation system is installed, the nozzles at first being over the blocks, but later moved to the slabs themselves. Once the plants have rooted into the slabs, slits are cut in the plastic wrapping to facilitate drainage. Several different ways of providing the necessary drainage are in use, including vertical or horizontal slits (5 cm long, 1 cm from the base) on the side away from the path, (if fully wrapped), or even holes made from underneath (Van Os 1982).

Static subirrigation has been adopted in The Netherlands, using rockwool as the substrate. The plastic troughs, made of polypropylene or polyester, are 5 cm high. In them stood rockwool blocks with a volume of 1–2 per plant. The solution should be about 2 cm deep for cucumbers and 3–4 cm for tomatoes, and the tops of the blocks should rise about 3–4 cm above water level. The plants, still in the smaller blocks in which they were propagated, stand on the supporting blocks in the troughs. A variant of this system has also been tried with partial circulation of solution. When solution is added periodically, a small amount is drained off to a collecting tank and may be reused. Van Os (1982) noted that both these forms of subirrigated rockwool culture were widely used for peppers in The Netherlands, and that tomatoes and cucumbers also grew well in such systems.

4.4.2 Nutrition and pH

The pH of new (unused) rockwool can be rather high (7–8), and values up to 9.0–9.5 have on occasion been recorded (Verwer 1976). This condition can be corrected before planting by using suitably acidified nutrient solution initially, for example pH 5.0–5.5; thereafter a slightly higher pH of 5.5–6.5 would be appropriate for many crops. The electrical conductivity of solutions within the slabs is maintained at 2.0–2.5 mS cm^{-1} for cucumbers and 2.5–3.0 mS/cm for tomatoes. These concentrations can generally be achieved in Holland by supplying nutrient solution at 1.5–2.0 mS cm^{-1} for cucumbers and 1.7–2.2 mS/cm for tomatoes (Sonneveld 1980).

Much consideration has been given in The Netherlands to the formulation of nutrient solutions for rockwool culture. Whilst such formulations are in many respects similar to those used in other forms of hydroponics, the concentrations of magnesium and iron often recommended for use with rockwool are relatively low. Thus Sonneveld (1980) proposed the following concentrations for cucumbers in rockwool, here recalculated in mg/l: 168 N, 47 P, 235 K, 18 Mg, 140 Ca, 0.56 Fe, 0.55 Mn, 0.26 Zn, 0.22 B, 0.05 Mo, and 0.03 Cu.

The distribution of soluble salts and pH within rockwool slabs has been studied by Van Noordwijk and Raats (1980); the measurements were made during midsummer when the tomato plants were in full production. Root surface area per unit volume of substrate was highest in the uppermost zone of the rockwool. The pH was also highest in the upper zone, particularly between the plants, where it reached up to 7.1–7.4. In contrast, the electrical conductivity was markedly higher in the bottom of the slabs (2.3–3.1 dS/m) as compared with 1.3–1.9 dS/m in the upper layer. These authors concluded that accumulation of excess salt was not uncommon in rockwool. The efficiency of leaching depended mainly on the relative positions of the drippers and drainage outlets.

Two stock solutions are usually made up to avoid precipitation in the concentrate storage tank which results when calcium- or magnesium-containing fertilizers are mixed with those containing sulfates or phosphates. These two stock solutions are diluted and combined in a mixing tank before application to the crop so that a complete nutrient solution is given at every watering (Table 4.2). The total concentration of nutrients in the irrigation water should be monitored on a continuous basis by a salt sensor (electrical conductivity cell). Similarly, the pH of the irrigation water, after all fertilizers and any acid have been added, should be monitored by a pH sensor meter and the rate of acid application be adjusted (automatic adjustment is available) to achieve a desirable pH for the nutrient solution. There can be many alternative feed recipes depending on the crop grown, the water source, the stage of crop growth, the season, etc.

A solution with a pH of 5 to 6 and an overall electrical conductivity (EC) between 2000 and 2500 micromhos/cm is generally recommended and is usually adequate for the crop. Nutrient changes based on crop appearance and growth will be necessary, especially in relation to seasonal variation of light availability. Daily checks of the pH and salt concentration (EC) of the slab solution are necessary, and bi-weekly analysis for all essential nutrients is highly recommended. Serious nutrient imbalances can be corrected by making appropriate changes to the nutrient formula, but these should deviate as little as possible from the normal solution. The general-purpose nutrient formula, as given above, should work reasonably well with both spring and fall crops, as well as with plants during the propagation stage. However, the total salt

Table 4.2. Stock solutions for rockwool-grown crops. (Ministry of Agriculture and Food, Ontario 1988)

	kg/1000 L
Stock solution A	
Calcium nitrate	64.0
Potassium nitrate	10.0
Ammonium nitrate	4.0
Iron chelate (13% Fe)	0.56
Stock solution B	
Potassium nitrate	30.0
Monopotassium phosphate	20.0
Potassium sulfate	4.5
Magnesium sulfate	18.5
Manganese sulfate	0.160
Zinc sulfate	0.110
Borax	0.180
Copper sulfate	0.012
Sodium molybdate	0.012

When these two stock solutions are diluted at 1:100, combined, and applied to a crop, they will supply all essential nutrients at the following concentrations in ppm:

Nitrogen	(N)	165	Manganese	(Mn)	0.4
Phosphorus	(P)	46	Zinc	(Zn)	0.25
Potassium	(K)	225	Boron	(B)	0.25
Calcium	(Ca)	122	Copper	(Cu)	0.03
Magnesium	(Mg)	18	Molybdenum	(Mo)	0.05
Iron	(Fe)	0.70			

concentration (EC) and the ratio between potassium and nitrogen should be varied according to the season and stage of growth. In general, higher ECs and potassium:notrogen ratios are recommended during the winter months when light is limited, whereas lower ECs and potassium:nitrogen ratios are recommended for summer months when the days are long and bright, and the crops use more water.

Some general recommendations based on commercial experience and research in England are summarized in Table 4.3.

4.4.3 Controlling the Water Content of the Substrate

A device for controlling the irrigation cycle, known as the Nieuwkoop pot, was included in Fig. 4.4b. This control device is connected to the rockwool slab by means of a short length of tubing inserted through the polythene wrapping. The tube contains a length of very thin glass-fiber material, rolled up to fit inside it; this restricts the rate at which water moves in and out of the pot and thus stabilies the control system. Irrigation commences whenever the level of solution falls. The bottom

Table 4.3. Target nutrient levels in NFT solution in ppm. (Beam et al. 1990, Ministry of Agriculture and Food, Ontario 1988)

pH		5.5	6.0	6.5
Conductivity (μS/cm)		1800	2000–2500	3500
		Minimum[a]	Optimum	Maximum
Nitrate nitrogen	(NO_3^--N)	50	150–200	300
Ammonium nitrogen	(NH_4^+-N)	5	10–15	20
Phosphorus	(P)	20	50	200
Potassium	(K)	100	300–500	800
Calcium	(Ca)	125	150–300	400
Magnesium	(Mg)	25	50	100
Iron	(Fe)	1.5	6	12
Magnesium	(Mn)	0.5	1	2.5
Copper	(Cu)	0.05	0.1	1
Zinc	(Zn)	0.05	0.5	2.5
Boron	(B)	0.1	0.3–0.5	1.5
Molybdenum	(Mo)	0.01	0.05	0.1
Sodium	(Na)	–	<30	<90
Chloride	(Cl)	–	<50	<150
Sulfur	(S)	–	50–200	–

[a]The concentrations listed as minimum are the approximate lower limit of a preferred range; in general, these minimum values are above those at which deficiency symptoms would develop

of the pot should be about 6 cm below the base of the rockwool slab (Welleman and Verbwer 1984).

In yet another system, a "control" slab of rockwool, not wrapped on the underside, is placed on a capillary mat, one end of which is curled down into a small container. Two electrodes are mounted over the container, one of which switches on the irrigation system when its tip is out of water, whilst the other terminates the irrigation cycle when the water level rises. a single-electrode system, known as the "start tray" is also available. This initiates an irrigation cycle when the electrode is "dry", the cycle afterwards being terminated by a pre-set timer. Various other ingenious devices have been tried, based on either weighing a sample slab (applicable to row crops where much of the crop weight is carried by the support wires) or on weighing the run-off from a slab.

As an alternative to monitoring the moisture status of the slabs, watering can be controlled indirectly by measurement of solar radiation. This reflects the water requirement of the crop during daylight hours, but leaves the plants without any irrigation at night despite the fact that some uptake of water continues through the hours of darkness. This problem can be overcome by incorporating a time clock to give three or four waterings during the night. Again, the whole irrigation program may be controlled by a timer, pre-set for maximum frequency of application at and shortly after noon, i.e., following the typical pattern of solar radiation and transpiration throughout the 24 h cycle.

4.4.4 Irrigation to Runoff

Excess solution drains quite readily from rockwool slabs, hence watering to excess can be extremely important in controlling nutrient concentrations in the root zone. Some nutrients are absorbed by plants more rapidly than others, and these (for example, nitrate and potassium) can become depleted whilst others (e.g., sodium and sulfate) accumulate. Furthermore, the composition of the nutrient solution supplied, reflecting not only the fertilizers added but also the mineral salts naturally present in the water supply, may not match the current requirements of the crop; indeed, it can at best only approximate to plant uptake.

Irrigation to excess in substrate which drains readily can indeed be highly advantageous. Thus, continued irrigation can maintain the concentration of nutrients which would otherwise become depleted, while displacing slats which would otherwise accumulate. Smith (1987) advocated 15–20% runoff under normal commercial conditions, or up to 30–35% if necessary to restore the nutrient balance in the substrate. Still higher rates of excess irrigation are frequently mentioned, and it may well be that the justification for these lies not only in control of nutrient concentration but also in maintenance of oxygen supply to the roots, since oxygen depleting can readily develop (see Sect. 3.4.3). Nitrite (NO_2) is frequently found to be present in the nutrient solution in commercial rockwool culture (Pickford et al. 1985), suggesting denitrification in an environment deprived of oxygen by the activities of other organisms utilizing root exudates. In practice, rockwool slabs should be irrigated whenever the water loss reaches 5–10% of capacity. This may require ten or more waterings per day for a rapidly growing crop (Smith 1987).

The runoff from "open" systems such as rockwool culture can, however, raise a potential problem, namely the contamination of water supplies; this is a topical and evocative subject. Nitrate in particular is regarded as highly undesirable in drinking water, and it is important that this and other nutrients do not enter the water supply. A target of 15% runoff over the course of the season has been proposed by the advisory service in the UK. Crum et al. (1985) noted this pollution problem with respect not only to nutrients but also to pesticides, and directed attention to recirculating systems for future development provided that the possibility of the spread of disease was overcome. It may also be mentioned that disposal of large quantities of used rockwool itself poses environmental problems; solution culture would avoid this difficulty.

To reduce costs, rockwool slabs can be used for more than one season provided they are effectively steam pasteurized between crops. A thorough flushing-out of accumulated salts with plain water is recommended before the slabs are pasteurized. They are then unwrapped, drained, stacked loosely in pallets and steam pasteurized for 1 h at 95 °C. Methyl bromide can also be used to pasteurize rockwool between crops, but steam pasteurization is more effective. After pasteurization, the slabs are rewrapped with polyethelene film and are again ready for use. Reused slabs do not need pH readjustment and are easier to rewet than new ones. Usually, they can only be reused once, as some breakdown in the fiber structure occurs with handling and pasteurization, and the airpore space in the slabs decreases with every reuse. Production costs may be reduced by use of recently introduced low-density (low-cost) rockwool slabs which are used for a single cropping season (Donnan 1988). As already mentioned on page 37, growing in polyurethane is similar to rockwool, but this growth media may be used for several years.

4.5 Gravel Culture – Closed System

A typical subirrigated gravel closed system of the early type is illustrated in Figs. 4.6, 4.7, and 4.8. The main components of such a system are (1) a growing bed filled with gravel, particle size of 1/2–3/4 cm, coarse sand, lava, tuff or any other similar inert medium, and with a subgrade piping system; (2) a supply tank or sump to contain the nutrient solution; (3) a pump to deliver the solution to the growing bed; and (4) a timing mechanism to actuate the pump. In these systems, the solution is customarily flooded into the bed to within about 2 cm of the surface of the medium and then drains by gravity back into the sump for reuse.

Such systems are capital-intensive because they require leakproof growing beds, as well as subgrade mechanical systems and nutrient storage tanks. Waterproofed concrete was used initially, and later fiberglass or heavy-gauge plastic film.

All beds should be level, and height variation should not exceed 10 cm. The storage tank must hold enough solution for simultaneous flooding of the beds.

Subirrigated systems of gravel culture are undoubtedly costly to build, and growers are not always prepared to commit their greenhouses to such seemingly permanent structures.

The beds require an accurate slope towards the outlet end, the gradients quoted in different sources ranging from 1 to 200 in about 1 in 1000. This slope is best achieved at the bottom of the bed, keeping the top level. Sloping the ground below the bed may be easier in construction, but it is then necessary to have a smaller depth of gravel at the upper end of the bed.

Various systems have been devised for circulation of the nutrient solution in gravel culture. In the direct feed system (Fig. 4.6), the solution is pumped from a storage tank into the drain at the base of the bed; the drain also acts as a feed pipe. When the solution reaches the required level in the beds, the pump is switched off and the solution drains back to the tank (Steiner 1976).

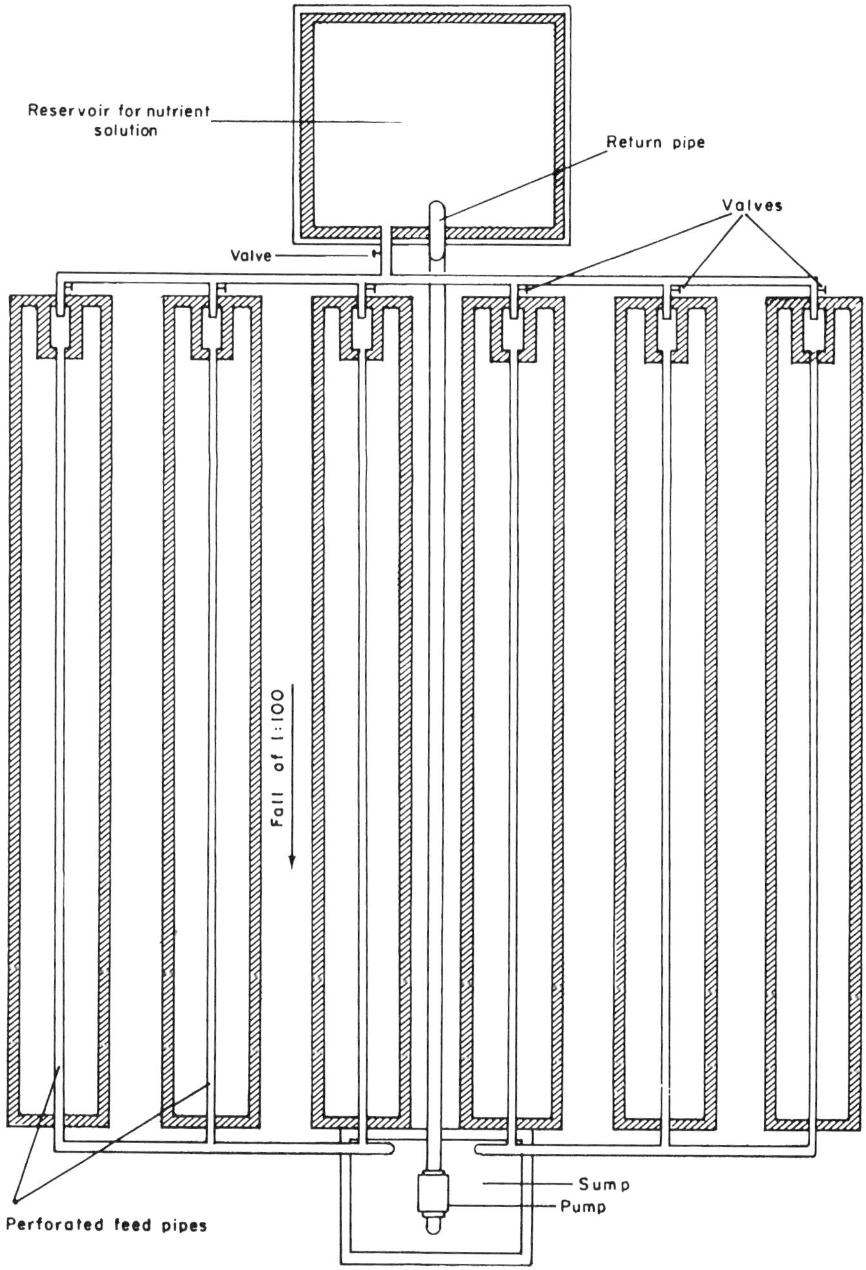

Fig. 4.7. Plan of a closed circuit system of subirrigated gravel culture with header tank for gravity feed (Stoughton 1969)

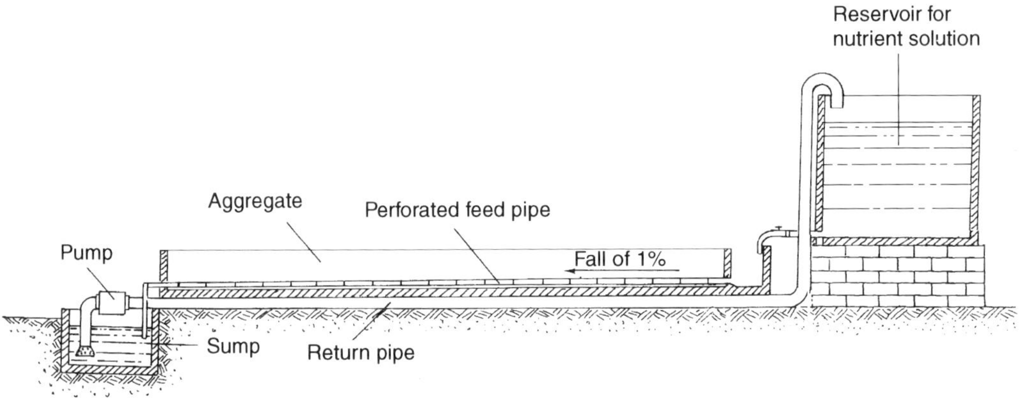

Fig. 4.8. Subirrigated gravel culture with gravity feed (side elevation) (Stoughton 1969)

This is referred to as the American system, as distinct from The Netherlands system, in which the beds have a flat base and the solution is made to flow through the aggregate during flooding. For short beds (up to 6 m) the direction of flow is along the bed, but the resistance is too great in longer beds. This problem is overcome by forming a duct along each side of the bed; solution enters on one side, flows across the bed, and leaves via the opposite duct. When the pump is stopped, the solution in the bed drains back into the tank via a small hole in the base of the bed, with a free fall to assist aeration.

Instead of direct pumping to the beds, a gravity feed system may be used, especially in large installations. Here the solution is pumped from the sump at the lower end of the system to a tank mounted aboveground at the upper end, from which it again runs into the beds. This allows smaller pumps to be used than in the "direct" system, since several hours are available, in which to return the solution from the lower to the upper tank. In a somewhat different form of gravity feed system, the use of pipes and valves is largely avoided, the solution being carried to and from the beds by a system of flumes, i.e., artificial water channels, generally formed of concrete. The beds are subirrigated by raising the level of the solution in the flumes and drained by lowering this level (Schwarz 1968).

The tank (sump) is usually built of concrete with its base sloping towards one side or corner. A small cistern has sometimes been incorporated at the lowest point in the tank, with the pump inlet sited within it to facilitate pumping out and cleaning.

The gravity-feed system has been adopted for use in some installations, taking advantage of any slight slope in the ground. The site is terraced (Fig. 4.9) a fall of about 2.5 cm is provided along each bed of approximately 20 m in length, with a fall of not less than 30 cm between sections. When the top beds have been filled, the solution is allowed to

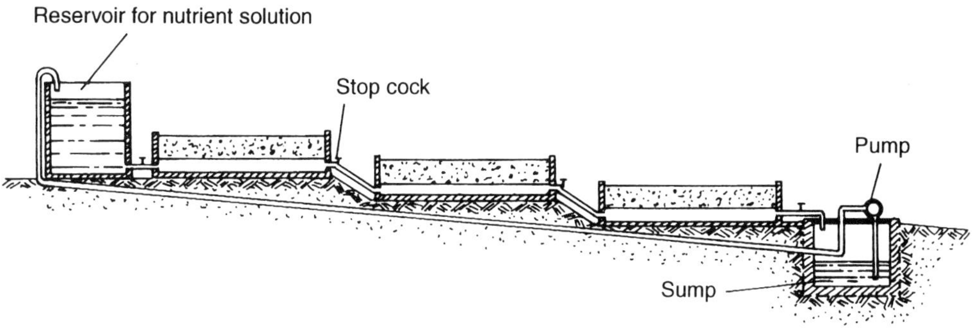

(a) DIAGRAMMATIC LAYOUT SHOWING ONE METHOD OF CONNECTING BEDS

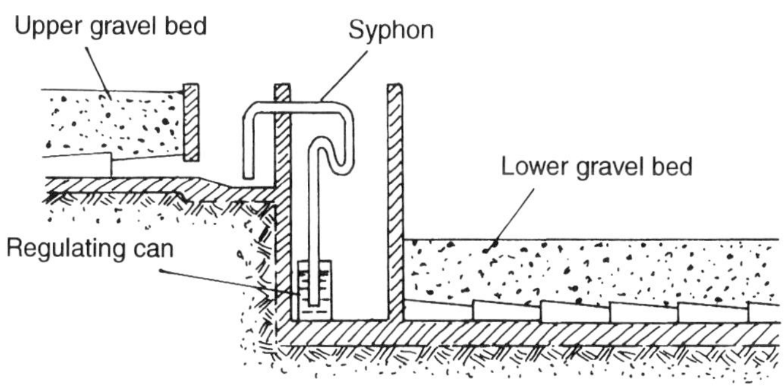

(b) ALTERNATIVE METHOD OF CONNECTION BY AUTOMATIC SYPHON

Fig. 4.9. The closed, subirrigated, cascade system of gravel culture on terraced ground (Stoughton 1969)

drain down to beds in the next terrace by means of either valves or of syphons. Since a proportion of the flowing solution is retained by each bed, the length is decreased progressively for those on the lower terraces in order to ensure fully adequate irrigation.

The main advantage of terracing lies in the smaller storage tank required – its capacity need be only about 30% more than needed for the top terrace, while the others are irrigated by gravity flow; five or six terraces with similar growing areas may be served by one tank.

An ingenious small-scale version of the terraced system, suitable for home use or possibly at remote locations can be seen in Fig. 4.10.

Each terrace is equipped with a cistern and ball (float) valve. Nutrient solution is pumped to the uppermost terrace; the water level rises and eventually begins to overflow into the cistern through a small outlet near the top of the bed. The float valve rises and opens the main drainage outlet from the base of the uppermost bed. The solution then overflows from the top cistern and fills the next bed in sequence. The lowest bed in the cascade drains into a catchment tank, from which it is pumped back to the uppermost terrace.

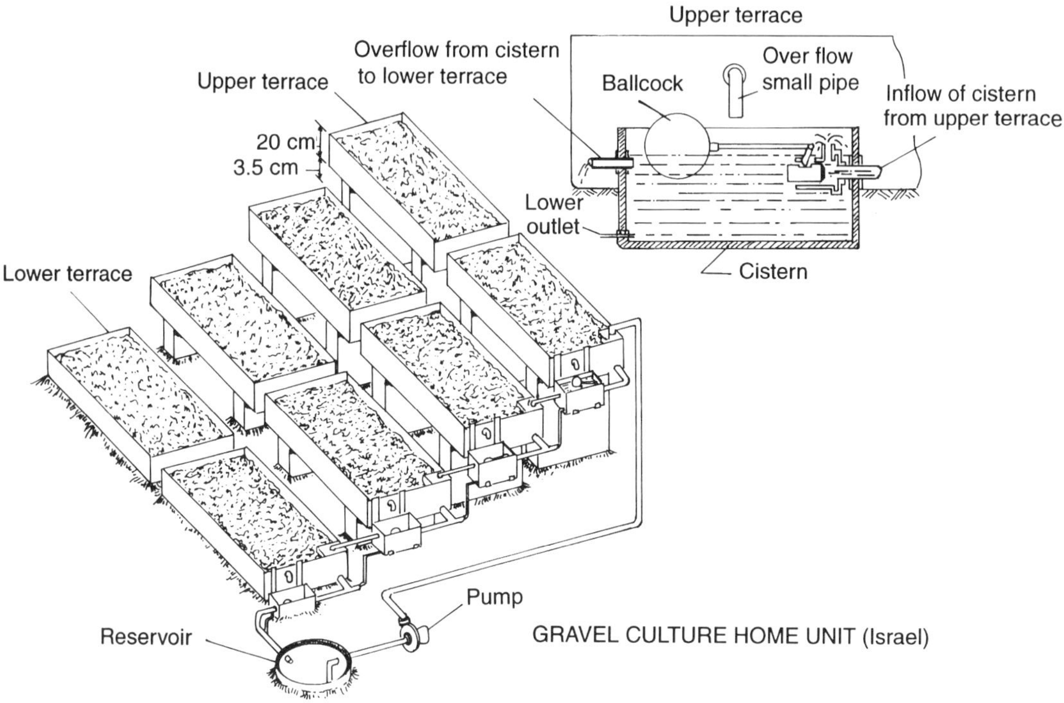

Fig. 4.10. Small-scale cascade system of gravel culture for home use in Israel (Schwarz 1968)

Static Subirrigation. Certain forms of soilless culture which do not fall clearly into either of the preceding sections are currently attracting considerable attention, particularly in Eastern Europe. Containers filled with a suitable substrate are stood in shallow troughs, to which nutrient solution is supplied to maintain the required water table. The commonest substrates are natural organic materials such as peat, wood bark, and mixtures of the two, though inorganic substrates such as perlite are also sometimes used in this way. In Poland, this method of crop production is known as the tray container technique or tct. The application of this technique is important in tomato production in an organic substrate consisting of pine bark and sphagnum peat (2:1 v/v). The plants are grown in specially shaped plastic containers with perforated sloping bases, standing in troughs containing nutrient solution (4–6 cm deep) into which the lower part of the container dips. The containers hold 7 l of substrate, sufficient for two tomato plants. The substrate is pretreated with fertilizer supplying N, P, K, Ca, Mg, and trace elements, and the irrigation water contains a proprietary liquid fertilizer. It was claimed that the system, which is simple to use, saved space in the growing houses by allowing the plants to remain longer at close spacing in the propagating unit. Circulation of the nutrient solution showed no advantage over static solutions.

4.5.1 General Layout

Farm size will depend on the crop grown. A farm covering 2000 m^2 net is generally considered economically worthwhile. In some circumstances, smaller farms may be practicable.

Three general types of layout are used in commercial gravel culture. The first, general for small farms where the topography does not allow terracing, consists of beds 1 m wide, about 30 m long, at intervals of 1 m. This layout may be used in farms reaching 1000 m^2. It provides easy access to beds; however, if the beds are to be covered with plastic or glass, spaces between them are best reduced to 40–50 cm.

In larger farms, if the terrain favors it, terracing is advisable; the nutrient solution is then pumped into the highest bed and flows by gravity to lower beds. Stony hillsides, unsuitable for conventional cultivation, may be put to good hydroponic use, with the advantage of protection from wind. While standard 1 by 30 m beds are generally used, beds may be longer or shorter according to the terrain.

The third type of farm uses large beds of 300–500 m^2 in area, without provision for cultivation paths. This layout may also be terraced. Heavy construction equipment may be needed to level the ground beneath such beds, which are generally asphalted, or covered by overlapping plastic sheets.

The advantage of this system is the lower capital cost per m^2 growing space. Maximum growth area is utilized – despite an approximate 20% loss for walking space – a special consideration for greenhouse crops. Levelling is generally done by heavy equipment, thus reducing manual labor. This layout is used where the plant growth media are inexpensive.

4.5.2 Bed Size

Once the general layout has been decided upon, suitable beds must be built. Beds 80–120 cm wide, 20 cm deep, and 15–30 m long are recommended for general use. Beds wider than 120 cm are difficult to cultivate; in beds less than 80 cm wide the growing space is not utilized efficiently (Fig. 4.11).

Beds 120 cm wide are generally used for flowers, and narrow beds 80–90 cm wide are used for two-row crops requiring trellises, such as tomato. Beds as narrow as 30–35 cm may be used for one-row crops, such as cucumber.

A 20 cm depth is generally sufficient for most crops. Deep-rooted crops such as roses and grapes need a bed depth of at least 25 cm.

Beds should slope about 1–2 cm per 15 m length (an approximate 1:1000 grade) toward the drainage end. A 2 cm solution level at the bottom of beds is not harmful for most crops. Thus, beds levelled inexactly,

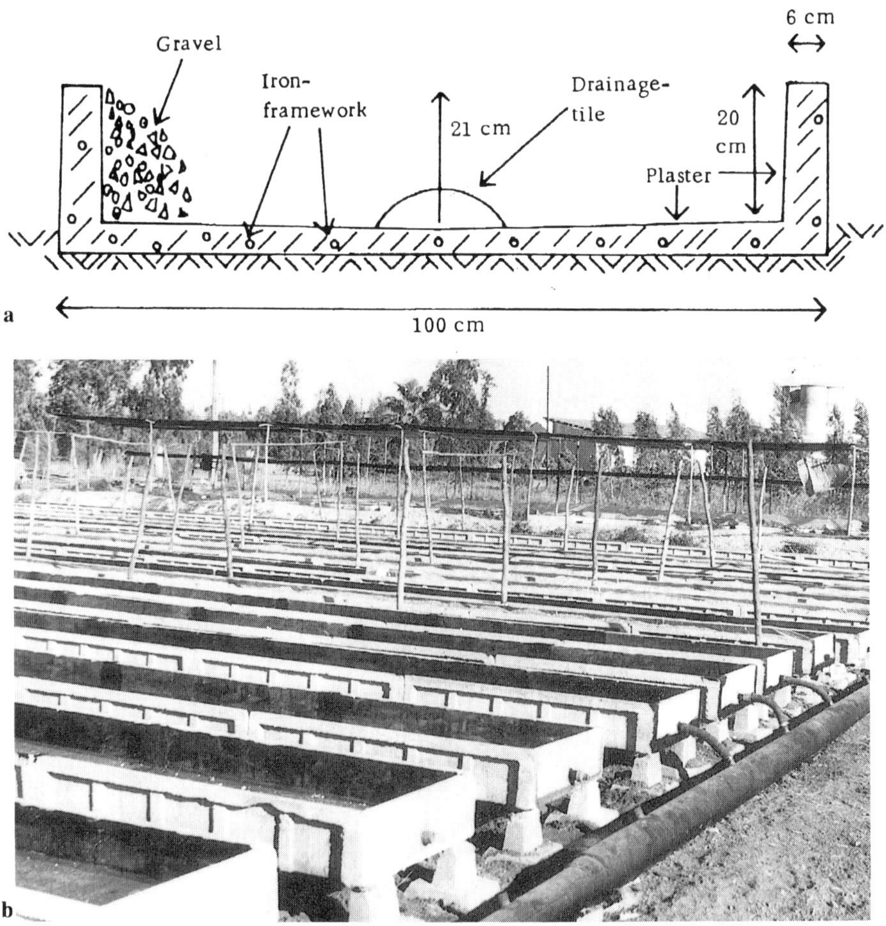

Fig. 4.11. **a** Concrete hydroponic bed (section). **b** Prefabricated concrete beds on supports, Hafetz-Hayim

which do not drain completely, may be used, provided that the solution level in no part exceeds 2 cm.

In the 250–500 m^2 beds in the third type of layout the length of any side should not exceed 25 m. The slope to the drainage end is as described.

4.5.3 Materials

1. Reinforced Concrete. This is the most satisfactory construction material for hydroponic beds. The wall width generally required is 6–10 cm; it renders the wall relatively expensive, but reinforced concrete is durable and has many other advantages. Bed floors can be constructed in V shapes to allow full drainage, although this is not usually necessary, and inlet pipes may be cast in position. Using concrete, construction of trellises is simple; 35 cm-long pipe sections (3/4") are imbedded to about 25 cm, every 3–4 m, leaving about 20 cm of pipe above wall level, to which other supporting pipes may then be connected. Provision for wires to support plastic bed coverings may also be made by leaving 1 cm-diameter holes every 60 cm. Concrete beds are usually built on the ground, so that the support medium's level is about 25 cm above the soil; they may also be built on concrete columns. Raised beds, which are easier to cultivate, are practicable only with concrete (Fig. 4.11b).

A number of precautions should be taken when reinforced concrete is used. Where temperature fluctuations are marked, beds should be poured in sections not longer than 10 m, with expansion strips laid between them – copper, lead, or asphalt-mastic being suitable.

Concrete is alkaline, while nutrient solutions are acid. To prevent interreactions and to avoid complication of the solution control and weakening of the concrete, the beds should be well insulated from the solution with a material like asphalt. Various coating materials are described in the next section.

Concrete is very expensive and has been replaced by less expensive plastic material.

2. Brick and Concrete. Brick walls with concrete floors may be practicable in some areas. After the concrete floors are poured, the bricks are laid and plastered with cement. As in other concrete constructions, expansion joints must be made if necessary, depending on temperature conditions. While cost per m^2 is about the same as for concrete, the work may be done by a layman. This type of construction is used for the open circuit system as illustrated in Fig. 4.12.

3. Earth Soil. Both material and construction work are cheap. Large beds are easily prepared with heavy equipment, but 100–120 cm-wide beds need skilled labor for adequate levelling. With some soil types, sand may be needed to create a smooth surface. Soil beds may be constructed easily by digging them about 15 cm deep and piling the removed soil into the paths, thus increasing the depth to about 25 cm. The beds' sides should slope 1:1 or 1:1.5; thus, if the bed is 90–100 cm wide at the floor, it should be about 100–115 cm wide at the surface. The length should not exceed 25 m. Paths at least 1 m wide should be left between beds.

Disadvantages of soil beds are the difficulty of determining the location of possible leaks and of repairing any breaks. Beds are deeper than, or level with, their surrounding paths, so that cultivation is more laborious than with other types of beds. Owing to the low bed level, soil may be blown into the beds; this may spread soil-borne diseases.

Soil beds are generally waterproofed to prevent leakage of nutrient solutions. A method in which the nutrient solution is allowed to drain

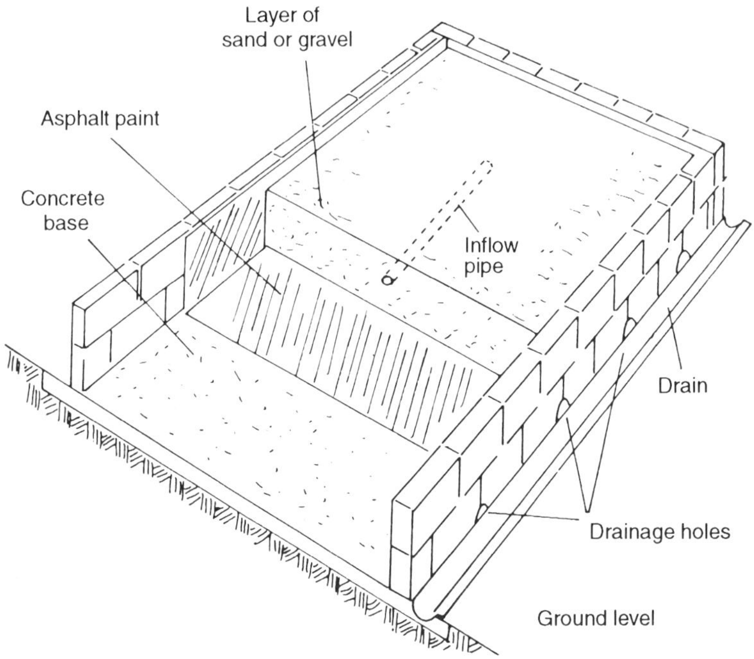

Fig. 4.12. Open circuit system in sand with overhead irrigation (FAO 1990)

into the ground is described in the section on sand culture. A cheap insulator is asphalt. Soil, or sand, must be completely dried to prevent the asphalt from subsequently becoming separated from the soil. One part of hot asphalt is mixed with two or three parts of separately warmed sand and poured while still hot on the floors and sides of the beds. A hot spade (best heated with a blowtorch) is used to distribute the mixture. Spreading is effectively performed with an iron bed-form connected to two parallel rails by a cross-bar rider, the rails being placed on the paths. A final thin-layer spray of pure asphalt is often necessary for a complete sealing.

Bitumen-coated material in 100–120 cm-wide rolls may be used successfully. Asphalt-saturated fiberglass sheets are also suitable. Since these sheets are usually not wide enough to cover a bed, they should be overlapped by 10–15 cm with hot asphalt. Asphalt-saturated fibrous materials, e.g., burlap, etc., may be used if locally available.

Upon heating, gravel-medium particles may penetrate asphalt membranes, interfering with drainage or breaking the membrane. Thin polyethylene plastic sheets (0.03–0.05 mm) will serve to protect asphalt membranes from such direct contact.

4. Plastic Sheeting. This relatively cheap material is often used to cover soil beds, mainly 0.2 mm polyethylene sheets or 0.15 mm PVC sheets.

Fig. 4.13. Gravel culture bed during construction. Side walls are composed of concrete channels. The inlaid folio is provisionally held in place by wooden slabs until it is filled with the substratum. On the bottom of the bed, a cable housing runs, mounting into a filter basket, shown behind an iron-tin case (*front*). A float in the latter automatically opens the drain as soon as the desired consistency of the nutrient solution is attained with stowaging (Institute for Soil Science and Plant Nutrition, Weihenstephan, Germany)

If the sheets are not available in the width required, they should be heat-welded with the proper equipment. Outlet pipes must be carefully connected, to avoid damage to the sheets. A hole is cut in the desired diameter and a pipe flange connected and firmly closed (Fig. 4.13).

Plastic sheets are useful for standard 1 × 30 m beds only (Fig. 4.14) and are not practical for large beds. Contact of nutrient solutions with most plastics causes no harm.

For large beds and open systems, polyethylene film at least 0.02 cm in thickness is now commonly used. The film, often in double layers to avoid leakage, (pinholes in either layer will seldom match up), is placed atop a sand base and supported by either planks, cables, concrete blocks, or the soil walls.

5. Coating. As previously noted, concrete beds must be lined with a chemically inert material such as asphalt before use. All prospective coating materials should be tested for toxicity. The best test method is by applying the material to a piece of filter paper and placing a few seeds on its surface, keeping the whole moist in a dish until the seeds germinate. The primary roots are then inspected with a magnifying glass and compared with the roots of seeds germinated on paper in similar

Fig. 4.14. Plastic-lined beds at soil level, filled with aggregate (coarse sand or gravel). Negev Institute, Beersheba

conditions but without coating. If those root tips and hairs in contact with the coating are shrivelled, the material is obviously toxic and should not be used. Coating materials should only be tested or used when completely dry because some may be toxic only when wet. Most bituminous materials are safe when completely dry.

4.6 Nutrient Film Technique (NFT) and Water Culture – Closed System

The NFT system, developed by Cooper (1975) in the UK, has been the focus of much attention over the past several years. The cost is still high, the disadvantages, especially root diseases, have not been overcome, and the economics are not yet clear. This system requires more scientific skill than the sand-rockwool-bag systems (Hurd and Price 1977; Graves and Hurd 1983; Benoit and Ceustermans 1986).

4.6.1 Nutrient Film Technique

The nutrient film technique is a novel system of water or solution culture characterized by using only a very shallow stream of solution flowing down the troughs or gullies. The plant roots form a more or less thin mat over the base of the gully, approximating to a two-dimensional rather than the usual three-dimensional root system. The solution is kept so shallow that the young plants, in their propagation blocks or pots, can stand in the gullies; the roots rapidly emerge into the flowing liquid. The high ratio of surface area to solution volume helps to ensure better aeration. The shallow layer of solution requires lightweight polythene sheeting and provides greater freedom to change the layout when required.

The basic features of an NFT system are as follows:

1. A catchment tank containing dilute nutrient solution.
2. A pump which delivers the nutrient solution to the upper ends of the gullies.
3. Parallel gullies in which the crop is grown, laid on a slope down which the circulating solution flows.
4. A catchment (or return) pipe, into which the gullies discharge, to conduct the solution back to the catchment tank.
5. Monitoring and control equipment to maintain the overall nutrient concentrations (total salts), pH, and water level. An electrical conductivity (EC) controller and a pH controller are commonly used to regulate the operation of pumps or valves controlling the transfer of fertilizers and acid from the supply tanks to the catchment tank. A constant water level in the catchment tank is easily obtained by a mechanical floating valve or by a variety of electronic controls.

The general layout is shown diagrammatically in Figs. 4.15 and 4.16.

A commercial tomato crop growing in NFT is illustrated in Fig. 4.20. Details of control system of NFT are described in the section on irrigation. The nutrient solution flows as a shallow stream, preferably less than 5 mm deep, at the rate of about 2 l/min down each trough.

4.6.2 Components of an NFT System

4.6.2.1 The Catchment Tank

The tank, being the lowest point in the system, is sited below ground level. It should be covered to exclude light and hence prevent algal growth.

Small systems can readily be provided with tanks large enough to contain all the solution from the system when drained back. Early commercial installations operated with a tank capacity representing only some 20–30% of the total solution volume, thus saving glasshouse space and construction costs. The full volume of solution present in the system is 20 000 l/ha at the start of the season, before the full root mat is established. Intermittent flow could be achieved with a smaller volume

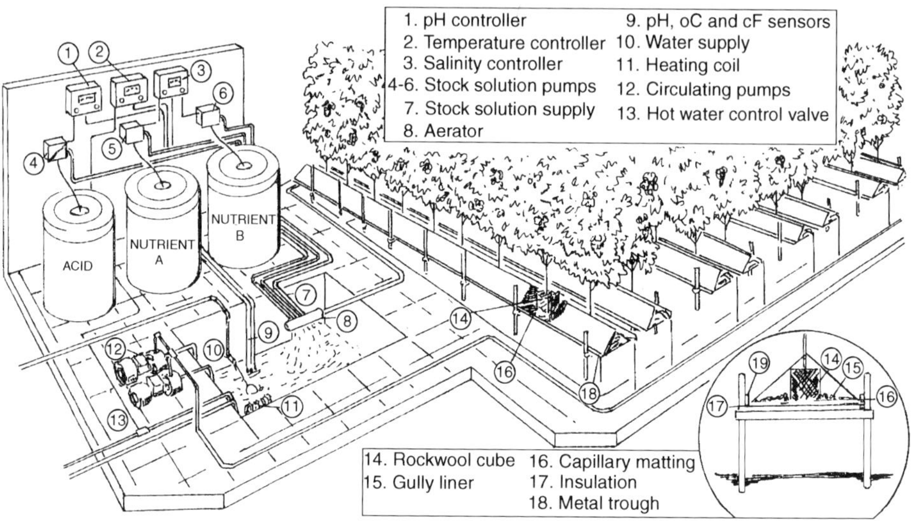

Fig. 4.15. General layout of an NFT system, showing gullies, control system, and catchment tank (Graves 1983)

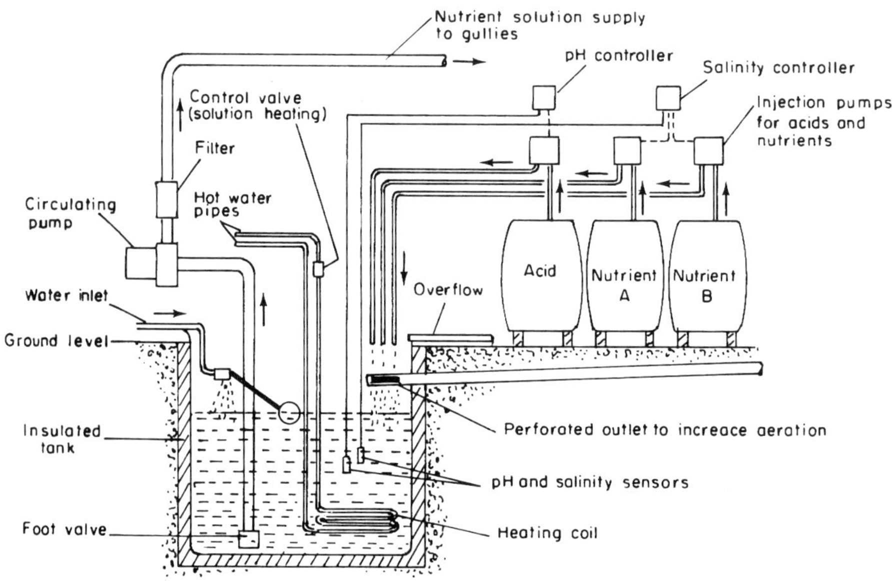

Fig. 4.16. The catchment tank and nutrient supply system in an automated nutrient film system. The main features are: *A* Foot valve; *B* insulated tank; *C* ground level; *D* water inlet; *E* circulating pump; *F* filter; *G* nutrient solution supply to gullies; *H* hot water pipe; *I* control valve (solution heating); *J* pH controller; *K* salinity controller; *L* injection pumps for acids and nutrients; *M* overflow; *N* perforated outlet to increase aeration; *P* pH and salinity sensors; *Q* heating coil; *R* return (catchment) pipe from gullies (FAO 1990)

of solution, circulated sequentially to different sections of the cropping system and controlled by solenoid valves. Management is undoubtedly simplified where the full tank capacity for the system is provided, and the problems of flooding and loss of nutrient solution resulting from pump failure or interruption of power supply is reduced. Provision should be made for the tank to be pumped out to waste when necessary.

Tanks fabricated from sheet plastic with metal reinforcement have proved satisfactory. Concrete tanks should be sealed inside with non-phototoxic epoxy resin to prevent contact with the mildly acid nutrient solutions. Pits lined with polythene film are a low-cost alternative, but cannot be recommended for commercial use; serious problems have arisen when water seeped into the pits behind the polythene sides of the enclosure. Problems have also arisen on occasion when rising water levels have forced empty or partially empty fiberglass tanks to float upwards in their pit. Domestic cold water tanks fabricated in fiberglass retention make effective and inexpensive catchment tanks for experimental systems containing up to 250 l of solution.

4.6.2.2 Circulating Pumps

The electricity-driven pumps should be of robust design for continuous operation and capable of pumping slightly corrosive dilute nutrient solution. Salt concentrations will normally be within the electrical conductivity range of 2–5 mS cm and pH range 5.5–6.7. Conductivities <8 mS cm are sometimes encountered for short periods (2–3 weeks), and pH values below the planned operating range may sometimes occur briefly. Two such pumps are normally provided, each capable of handling the pumping load. Should the operational pump fail, the second pump is switched in.

Some early descriptions of NFT systems included a header tank, fed from the catchment tank by the circulating pump and incorporating a submerged float valve to admit mains water in the event of pump failure. Nutrient solution flowed under gravity from the header tank to the upper ends of the gullies.

4.6.2.3 The Gullies and Plant Supporters

Although early NFT systems used a galvanized iron wire above the center of the gully to support the edges of the polythene, this is not necessary for crops propagated in pots or in rockwool cubes (7.5 cm or larger). The pots themselves suffice to support the plastic, the sides of the gully simply being clipped or stapled together at intervals.

A wide flat base is required to accommodate the expanding root system; narrow gullies retard the flow of the solution through the root mat and thus increase the depth of liquid and restrict aeration. A width of 30 cm is recommended for tomatoes and cucumbers. Despite every attempt to maintain the level across the gullies, difficulty may be experienced when the plants are first set out, since the flowing solution may veer to one side or the other and thus fail initially to wet the plant pot or block. This problem will disappear when the plant roots emerge from the blocks and form a mat between them. For the early stages of growth, however, it is advisable to place narrow (5 cm) strips of capillary matting across the gullies beneath each pot or block; this will ensure that every plant will be watered adequately from the start. The plant roots grow down through the matting to proliferate underneath, thus decreasing gaseous exchange with the atmosphere.

The gully layout slope down must be sufficient to ensure a ready flow from top to bottom. This improves aeration in the flowing solution by increasing the ratio of surface area to volume, and prevents the water table from rising too high in the pots or blocks. A slope of 1 in 75 (or 1.33%) is recommended and never less than 1 in 100 (1%) or greater than 1 in 50.

The early NFT systems were laid on carefully sloped soil. It proved difficult to provide the accurate slope required for successful operation in this simple way. Sloped concrete floors proved very satisfactory, though more expensive. Furthermore, the sloped concrete floor is a permanent commitment, keeping out the flexibility normally associated with NFT systems.

The gullies should always be kept closed to exclude light, to prevent the growth of algae; short additional pieces of polythene sheet may be used for this purpose at the ends of the rows. Long gullies accentuate problems of poor aeration; a maximum length of 20 m is suggested, with shorter lengths (10–15 m) if possible.

4.6.2.4 The Catchment (Return) Pipe

A plastic pipe is laid at right angles to the gullies at their lower ends. Connection between the polythene gullies and the catchment pipe is normally by a short length of flexible plastic tubing. The return pipe is normally beneath the soil surface and is sloped downwards towards the catchment tank. Use of an open polythene-lined trench is not recommended.

4.6.3 Special Forms of NFT

4.6.3.1 Lettuce in Concrete Gullies

Conventional gullies of polythene, devised for use with row crops such as tomatoes or cucumbers, are obviously unsuitable for lettuce on a commercial scale. Lettuce needs to be planted at close spacing (20 cm × 20 cm) with a minimum of path space. The rosette form of the plant is not suitable for the customary closed gully of triangular cross section. The lettuce is grown in casting shallow gullies, approximately 10 cm wide, 2 cm deep and 10 cm apart, in a sloped concrete floor of the glasshouses (see Fig. 4.11). The gullies run the full length of the glasshouses.

The concrete is treated with an epoxy resin to prevent chemical attack, and can be driven over by battery-powered rubber-tired vehicles for planting and harvesting; these three-wheeled vehicles are self-steering, the wheels running along the sunken gullies.

No paths are provided within the cropped area, and the gullies have no form of covering to restrict the movement of either staff or vehicles. Some algal growth develops in the exposed gullies during cropping, but this is not considered detrimental to this short-term, close-planted crop. Nutrient solution is delivered to the upper ends of the gullies via rigid PVC pipes, laid at right angles to the gullies with outlets to each line of plants. Conductivity and pH controllers, catchment tanks, and circulating pumps are retained.

4.6.3.2 The IMAG System for Lettuce in Alternation with Tomatoes

Problems arise where the cropping program requires alternation of row crops with lettuce, as is often the case where only limited heating capacity is available. While it would be possible to remove the NFT troughs at the end of the tomato season and plant lettuce in the underlying soil, this would be both laborious and expensive, since soil sterilization would still be required.

Another system was devised at the Institute of Agricultural Engineering (IMAG) at Wageningen (van Os and Kuiken 1984). Thus, for lettuce, seven parallel lines of troughs cover the floor of a 3.2 m-span greenhouse, and four of these are used for the following tomato crop, the other three being moved aside to create paths. The special features of this system relate mainly to the lettuce crop, the tomatoes being grown by standard procedures.

The surface of the soil is graded to a slope of 0.5–1.0, the flow and return pipes for irrigation are installed and the surface is covered with white plastic sheets. The black polypropylene troughs, 25 cm wide and 5 cm deep, are then rolled out in lengths of up to 60 m. The sides of the troughs are raised and their lower ends connected to the return (catchment) pipe. Nutrient solution is conveyed to the upper ends of the troughs by small (3 mm diameter) pipes, used in pairs, the irrigation rate being about 3 $l/m^2/h$. Special lids are provided, made of white polypropylene (30 cm wide). A prototype machine was developed by IMAG to cut out square holes (5 × 5 cm) in the covers. Two parallel rows of holes are made along each cover, the rows being 20 cm apart with holes at intervals of 22 cm. The lettuce are propagated in planting blocks of peat or other material, 5 cm high and with a 4 × 4 cm base.

The machine propels itself forward by means of an electrically driven winch and cable. Two operators travel on the machine to place the planting blocks in the holes as the cover is unwound.

This highly mechanized system has been tested in some commercial nurseries. Published costings, taking the cost of soil sterilization and the more rapid crop "turn-around" into account, suggest that the IMAG system could be economically viable. It was envisaged that the capital cost of the harvesting machine would be shared between several nurseries.

4.6.3.3 Further Systems for Lettuce Production in NFT

A mechanized system for continuous lettuce production, developed by Whittaker Agri-Systems (California), has been described in outline by Rogers (1983). The lettuce are set in plastic cups, supported in single lines by a carrier strip above a fixed trough of circulating nutrient solution. The plants are arranged in two tiers so as to make the most effective use of greenhouse space. At harvest, a machine at the end of the house winds up the carrier strip and pulls the plants from the trough past a knife which separates the heads from the roots. At the same time, a further carrier strip with young plants is drawn into position over the troughs.

Another scheme for production of greenhouse lettuce at high planting densities were developed by Morgan and Tan (1983). The plants are located in 2 cm holes cut at 20 cm intervals in plastic tubes (3 cm diameter), these being laid in tiers on an archway pattern. The growing tubes, each 6 m long, are arranged at a slope of 1 in 30 for effective circulation of solution. Up to 12 tiers can be accommodated from top to bottom, though recent results suggest that six double tiers might be preferable. The young, bare-rooted plants are inserted into the holes, allowing the leaves to rest on the upper surface of the tube. Planting densities of 40 plants/m^2 are obtainable. After each crop, the roots are pulled or pushed out of the tubes, and the tubes are sterilized by immersion in formaldehyde for 10 min.

4.6.3.4 Strawberries in NFT

In Belgium and The Netherlands, some 5–10 ha of strawberries are currently grown in this way. The planting density is approximately 13.5 plants/m^2. The troughs are supported from overhead, and all troughs are normally at the same height (80 cm). For working purposes, however, two or three troughs are raised to a height of about 2 m so as to create paths for the staff; the troughs have long flexible connections for this purpose.

The troughs used for this crop are relatively narrow, being only 6.5 cm wide (Van Os 1982). The presence of propagating blocks in such narrow gullies would unduly restrict the flow of nutrient solution. Bare-rooted plants are therefore used, and special holding devices are accordingly designed to support the cuttings throughout the whole production cycle.

The cuttings, in their special holding device, are first rooted in a controlled environment chamber at high humidity. The holders are then slid into the growing trough on a transportable bench and moved to a growth room under long day conditions for early vegetative growth. The benches are next moved to a second growth room under short days so as to initiate bud formation. When the blooms are ready for cutting, the benches are moved to a working area for mechanical harvesting, grading, and bunching. The system is so arranged that, at all stages of production, a flow of nutrient solution is maintained down the 1% slope of the troughs.

For chrysanthemum production, gullies are created by fixing expanded polystyrene strips 2.5 cm wide and 3.8 cm high on to a basal slab of expanded polystyrene. The strips are arranged in parallel lines, 10 cm apart, and covered with black polythene sheeting. The beds, approximately 1.5 cm wide and 18 m long, are finally covered with white-on-black polythene, through which the chrysanthemums are "planted". The cuttings are rooted in vertically grained rockwool cubes, pretreated with an acidic solution to lower the pH of the material.

4.6.4 Aeration in NFT Systems

The decreases in oxygen content from entry to outlet for tomatoes were from 8.3 mgO_2/l at the inlet to 7.5 mg/l at the outlet, and in cucumbers from 6.2 mgO_2/l at the inlet to 2.9 mg/l at the far end. Values below about 3 mgO_2/l in the solution are regarded as undesirable.

Depletion of oxygen levels in the solutions used for cucumbers doubtless reflects the large root system produced by this crop.

Values as low as 1 mgO_2/l resulted in wilting of the plants on sunny days.

It is, of course, not only the concentration of oxygen in the recirculating solution that is important for the well-being of the crop; the roots must also pass through the containers or blocks in which the plants were propagated, and these can readily become waterlogged. As shown by Jackson et al. (1984), the solution within the propagation substrate can indeed be very poorly aerated.

Thus a peat-based compost in "whalehide" pots placed in NFT gullies was found to be severely waterlogged, with oxygen levels down to 0.7 mgO_2/l and evidence of denitrification. The use of more open-structured inorganic substrates such as rockwool or perlite was therefore recommended.

Various means of improving the aeration of solutions returning to the tank are in use, involving free-fall with splashing; the point of return should be well above the surface of the solution. Slatted plastic piping can be used as an alternative to an open-ended return pipe in order to increase splashing and foaming. Where there is more than one return pipe to the tank, it can sometimes be arranged for the streams of solution to converge at speed well above the level in the tank, again improving aeration. The most effective technique appears to be positive oxygenation by fitting a venturi-type aerator in a bypass from the pump, returning a continuous stream of oxygenated solution to the tank. The importance of such devices is widely recognized in Japanese soilless culture.

One disadvantage in having submerged roots is that they may not have sufficient available oxygen. Oxygen is relatively insoluble in water, its concentration at saturation decreasing with increasing solution temperature (15 °C = 10.2 ppm, 25 °C = 8.5 ppm, 35 °C = 7.1 ppm, 45 °C = 6.0 ppm). Higher temperatures also increase the respiratory demand (maintenance respiration) for oxygen by the roots; this approximately doubles for each 10 °C rise in temperature to about 30 °C (Schwarz and Gale 1983b). It was calculated that a mature tomato plant root system needs about 20 ml oxygen/h, so that with a solution rate of 1–2 l/min, the oxygen in solution would be depleted by 20–40 plants. However, this assumes that all the roots were submerged and that the nutrient solution absorbed to aerial oxygen (Hurd and Price 1977). In NFT systems, roots are only partly submerged in the solution, depending on the slope and width of the gully.

Oxygen is also absorbed as the solution passes down the gully, to an extent depending on flow rate and turbulence.

In NFT systems, the oxygen supply to the submerged roots may not be uniform. Once a mat of root has been established, free movement within the mat is restricted and the bulk of water flows along its edge, leading to pools of stagnant water within the mat. Jackson (1980) showed that the more deeply the roots are submerged, the more effective is the barrier to gas movement between the roots and the aerial environment. Jackson (1980) calculated that at 10 °C the oxygen dissolved in the nutrient solution (2 l/plant) which is available to a tomato plant at first anthesis, with a non-sterile root system, would require replacement twice each hour. At 20 °C it would need replacing five times each hour, and at 30 °C twelve times.

A root system in anaerobic conditions results in slow growth, senescence and abscission of leaves, petiole epinasty, and adventitious rooting of stems, all characteristics of waterlogged plants and associated with the production of ethylene by the plant (Jackson 1980).

4.6.5 Root Death and Wilting in NFT Culture

Poor growth and fruiting (including wilting of leaves and low yield of middle trusses in tomato, and, in severe cases, wilting and death) have frequently been observed in various NFT culture systems. However, in cases where the plants were retained, they eventually recovered, and later cropped relatively normally.

These observations were associated with death of the roots, resulting mainly from poor aeration. the normal root mat has pale brown roots, some dead roots, and, when recovery occurs, white roots grow over the surface of the root mat. The lack of oxygen is not the only cause for root death, as it occurs even in aerated NFT conditions. A competition for carbohydrates in the fruit-developing stage for the roots is suspected. Any toxic substance releases to the roots could also be the reason. In any case, the problem has not yet been solved.

4.6.6 Nutrition in NFT Systems

There is a basic difference in the nutrient concentrations necessary to substain growth in static and flowing systems of solution culture.

Concentrations which have proved highly satisfactory for sand culture and for static solution culture are far higher than those found necessary by research workers using flowing nutrient solutions (Winsor 1982).

In practice, however, it seems highly desirable to work at higher concentrations in commercial NFT systems in order (1) to ensure a reserve of nutrients within the system at all times, and (2) to permit control of nutritional levels by measurement of electrical conductivity. Since most irrigation waters contain an appreciable (and sometimes very high) concentration of salts, it would be impracticable to monitor low nutrient concentrations against an accumulating background of calcium, sulfate, or chloride such as commonly occurs.

A typical nutrient solution suitable for tomato growing in NFT would approximate to the following composition, expressed in mg/l: 150–225 N, 30–45 P, 300–500 K, 150–300 Ca, 40–5 Mg, 3–6 Fe, 0.5–1.0 Mn, 0.1 Cu, 0.1 Zn, 0.3–0.4 B, and 0.05 Mo. Higher concentrations of the major nutrients can, with advantage, be used to control early growth of tomatoes under conditions of poor light. For other crops, however, a somewhat lower K/N ratio would normally suffice.

Fertilizers and acid are normally pumped into the catchment tank as concentrated (stock) solutions. The size of these pumps (two for fertilizer and one for acid) depends upon the size of the operation, but an average capacity of 10 l/h is what most growers will need. Complete nutrient- and acid-dosing sets are now available as ready-to-use packages from a number of suppliers, although tailor-made systems to suit individual needs can be assembled with relative ease by the grower himself, as most of the needed components are widely available. In addition to the three pumps, a salinity (EC) and pH controller are also required. It is advisable to use a number of smaller pumps instead of a single large one so that some solution flow is maintained even if one pump fails.

To guard against total power failure, a stand-by generator is essential, especially for large installations.

Where the water supply contains nutrients in excess of plant requirements, or the fertilizers are not supplied in a ratio proportional to nutrient uptake by the crop grown, a buildup of certain nutrients will inevitably result and cause the "background" conductivity to progressively rise. Nutrients that can accumulate over time include calcium (from hard water), sulfate (from fertilizers), sodium and chloride (from saline water), and possibly others. This potential problem dictates that the nutrient solution should be discarded periodically and a new one made up. The frequency of renewing the nutrient solution will depend on the stage of crop growth and the season, since both affect the rate of nutrient and water uptake by the crop. A general recommendation is to renew the solution every month at the beginning of a crop and twice a month later when the crop is fully grown, or whenever the crop appears to have stopped growing. As one's experience with NFT increases, less-frequent solution renewal may be practiced. Frequent (weekly) chemical analyses of the nutrient solution in the initial stages of establishing a new NFT operation are essential for crop safety and grower education. Less-frequent analyses of the solution (twice a month) are needed later.

The pH of the nutrient solution also has considerable influence on crop growth, and needs constant monitoring and control. A pH range of 5.5–6.5 is considered optimal for most crop plants. Values below 5 are to be avoided and a pH below 4 will damage most crops. The availability of some microelements decreases when the pH rises above about 6.5.

Five to ten ppm is preferable during the early stages of growth before the start of fruit picking (Graves 1983).

Addition of chemicals to NFT systems may be required daily or weekly after either conductivity measurements or, preferably, chemical analysis for major nutrients (Table 4.4). In automatically controlled systems it is usual to prepare only two stock solutions, one containing calcium nitrate alone, and the other the remaining chemicals, to prevent precipitation of sparingly soluble calcium salts.

In automatically controlled NFT systems, the salinity controller is usually set at 2500–3500 μmhos depending on the "hardness" of the water.

The best-known improvement in NFT systems is the Ein Gedi (Israel) system (Soffer and Levinger 1980). This system was not found economical. The Ein Gedi system is a flowing nutrient solution culture system in which plant roots are completely immersed in a deep, circulating, continuously aerated nutrient solution.

Plant performances on the system are achieved by the combination of the following features and principles in one system:

1. The root system is thoroughly immersed in a large volume of nutrient solution which provided luxurious feeding and watering.

Table 4.4. Nutrient concentrations and chemicals for tomatoes in NFT

Element	Desirable concentration (ppm)	Chemicals
Nitrate nitrogen	150–200	KNO_3, NH_4NO_3, $Ca(NO_3)_2$
Ammonium nitrogen	0–20	NH_4NO_3, $(NH_4)_2\ SO_4$
Potassium	300–500	KNO_3, K_2SO_4, KH_2PO_4
Phosphorus	50	KH_2PO_4, NaH_2PO_4, $CaHPO_4$
Calcium	150–300	$Ca(NO_3)_2$, $CaSO_4$, $CaHPO_4$
Magnesium	50	$MgSO_4$, $Mg(NO_3)_2$
Iron	3	FeEDTA, FeEDDHA
Manganese	1	$MnSO_4$
Copper	0.1	$CuSO_4$
Zinc	0.1	$ZnSO_4$
Boron	0.3–0.5	H_3BO_3
Molybdenum	0.05	$(NH_4)_6MO_7O_{24}$
Sodium	Maximum 250	
Chlorine	Maximum 200	

2. The nutrient solution is sprayed and enriches the plants' oxygen concentration.

The flow rate is not less than 7 l/plant/h and requires 7 l of solution per tomato plant (two plants per m^2). (A fully grown tomato plant requires 1.5 l/day in England, and 2.5 in May in Israel).

NFT and Rockwool: A more recent development in Europe is the combination of NFT and rockwool culture, which is, by definition, a closed aggregate system. In this application, which is not widespread, plants are established on small rockwool slabs positioned in channels containing recycled nutrient solution (Hanger 1982).

4.7 Water Culture – Closed System

The starting point to grow plants without any soil was to replace the soil completely by a nutrient solution, meaning water, in which all growth conditions had to be fulfilled. The water culture, in its most simple form, consists of a glass with the nutrient solution in which the roots of the plant are suspended, and a support for the stem of the plant. This method was tried around the year 1665 by the famous Irish scientist, Robert Boyle.

Since the mid-19th century, the water culture method has been used in the study of plant nutrition. By dissolving different salts in the nutrient solution, the kind of elements a plant needs for its growth and development have been determined. In this way the water culture method has been the spring-board for the use of mineral fertilizers in agriculture and horticulture. The water culture method is still being used for the study of plant nutrition.

It was in 1929 that Prof. Gericke in the USA introduced the water culture method "commercially" for the production of tomatoes (Gericke 1937).

4.7.1 Deep Water Culture (Old Soilless Culture System)

A system of troughs (iron, concrete, plastic, etc.) of 50–80 cm width, 15–20 cm depth and 1–10 m in length holds the nutrient solution. Wire netting is stretched over the top of the trough on which are placed layers of sawdust, straw, or peat moss, 1–3 cm in depth.

When seeds are sown in the litter bed, the level of solution in the tank is raised to the bottom of the wire grid, thus watering the substrate by capillarity. When the roots enter the solution beneath, the water level is lowered.

An air space, increasing with the growth of the plants, is left between the covering of the tank on which the plants are supported and the surface of the solution. Often two kinds of roots are produced – absorbing roots growing in the solution and aerating roots in the saturated air space above the surface. The roots formed in moist air within the gullies are densely covered with root hairs, whilst those formed below the surface of the solution are far less so. Some growers have attempted to overcome the problem of poor aeration by continuous circulation of the solution or "splashing" the solution.

This original hydroponic system of Gericke's was thoroughly tested in several countries and generally found to be unsatisfactory in practice, mainly due to poor aeration in the solution. Furthermore, the difficulty and expense of establishing and supporting the plants above the solution made the system non-commercial.

4.7.2 Floating Hydroponic Systems

Floating hydroponics is a form of water culture in which the plants are supported above the surface of the solution on "rafts" of lightweight plastic material, expanded polystyrene being the obvious choice.

An installation of this type was described by Massantini (1976): the bed, 1 m wide, 3 m long, and 15 cm deep, was made of timber lined with plastic film, and the floating panels were 1 m square and 2 cm thick. The nutrient solution was recirculated, with aeration controlled by a timer. The crops are located in 15 mm-diameter holes at suitable spacings in the supporting rafts. The investment and running cost is less than conventional old-fashioned hydroponic systems (e.g., subirrigated gravel culture). Yields per plant were nearly always inferior to those obtained in soil, but yield per unit area could be increased by using high planting densities.

In installations in Arizona (Collins and Jensen 1983), lettuce and other leafy crops were supported on expanded plastic material (2.5 cm thick), floating on water (0.3 m deep) in troughs ("raceways") 70 m long and 4 m wide. Conveyer belts for planting and harvesting on moveable floats can be added. The floats can be harvested at one end of the tank, while new floats with transplants are introduced at the other end. Aeration problems are similar to conventional water culture. The economics of the system seem discouraging.

4.7.3 Japanese Systems

A wide range of beds or troughs is in use, including both pre-fabricated units and those constructed on site from concrete blocks covered with plastic sheet. The techniques most commonly used are described below (Suzuki et al. 1984; Fig. 4.17).

The Kyowa-system Water Culture Hyponica is described as a "semi-deep" system used mainly for vegetables. In this system the nutrient solution is pumped from a storage tank via an "air mixer" to the cropping bed, from which it returns underground to the tank via an overflow pipe. A cross section of the beds is shown in Fig. 4.17a. The beds are made up from units molded in rigid ABS plastic, each unit being 1 m wide (0.9 m internally) and 3.15 m in length. The plants are rooted in plastic pots with openings in the sides and base, supported by the cover over the bed.

The M-system dispenses with the use of any storage (catchment) tank. Nutrient solution in the beds is drawn off by a circulating pump; flows through an air mixer and is then fed back into the beds through small holes in pipes situated along the bottom. A cross section of the beds is show in Fig. 4.17b. The beds are constructed in expanded polystyrene, 0.66 m wide (0.6 m internally), using units 1.2 m in length to form beds 20 m long. To prevent leakage at the joints and to protect the polystyrene, the inside of the beds is covered with thin polythene film.

The Kubota system uses molded rigid ABS plastic. Individual units are 3.25 m long and 0.7 m wide (0.6 m internally), subdivided as shown in Fig. 4.17c. In the Kamizono system (Fig. 4.17d) the beds are of a simple type built as for traditional aggregate culture. The side walls are constructed from concrete blocks, and the beds are lined with thick black polyethlene sheet which lasts for 5–6 years. The black sheeting is in turn covered with a thin (0.1 mm) polyethylene film which is replaced each year, thus avoiding the need for sterilization. Yet another system (Shinwa) operated on the novel principle of periodic exchange of nutrient solution between pairs of equal-sized beds; the water level in each bed alternates between high and low, thus improving aeration.

Circulating pumps are activated for periods of 10–20 minutes in each 1–2-h cycle, according to crop requirement, with short periods at low

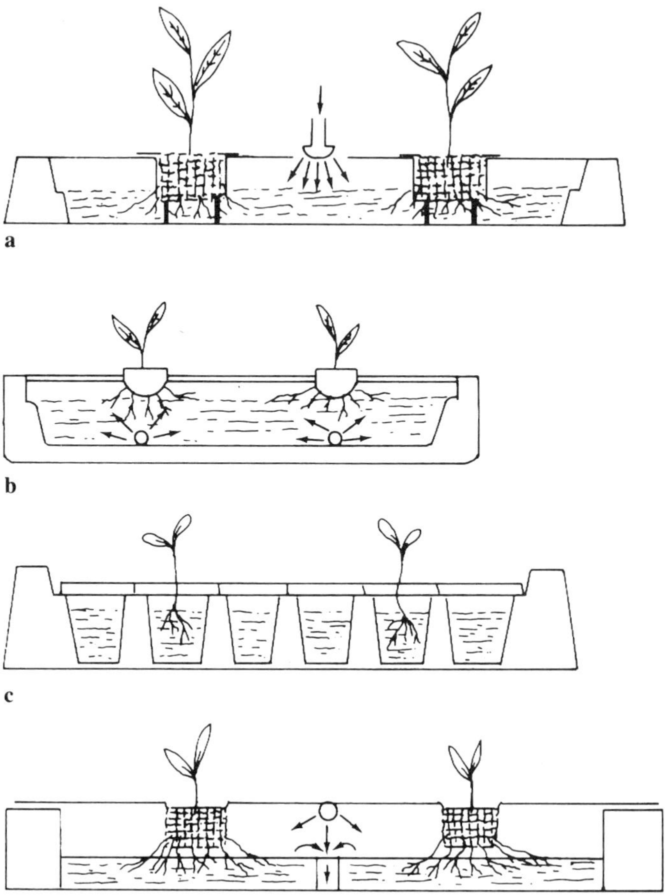

a b c d

Fig. 4.17. Some forms of deep or semideep recirculating solution culture used in Japan, here shown in cross section (FAO 1990). **a** Kyowa hyponica. Molded beds made of rigid ABS resin. Solution is supplied to the beds via an air mixer. **b** The M-system. Constructed from expanded polystyrene units and incorporating an air mixer. **c** The Kubota system. Molded in rigid ABS plastic with internal subdivisions. **d** The Kamizono system. A traditional type of hydroponics bed with side walls of concrete blocks and a polythene lining

temperatures and extended periods during hot weather. The pumps are operated once or twice during the night. For economy of pump size, large hydroponics units may be subdivided into blocks and fed sequentially.

Bed covers are used to support the plants, to exclude light and dust, and to avoid drastic changes in temperature within the beds. For row crops such as tomatoes, cucumbers, and melons the covers are of expanded polystyrene (1–2 cm thick), with holes cut to support slatted pots through which the plant roots can grow down into the solution beneath. For leafy vegetables, the polystyrene covers are manufactured with holes at a spacing of 10 × 7 cm. Gravel or plastic foam is often used as the substrate in the pots.

4.7.4 Aeroponics I (Nutrient Mist)

Steiner (1976), defined aeroponics as a system in which "... the roots of the plants are continuously or discontinuously in an environment saturated with fine drops (a mist or aerosol) of nutrient solution". The concept, which does not involve a continuous liquid phase, has sometimes been used in nutritional research, particularly with young fruits.

Aeroponic systems with potential for commercial use were described by Massantini (1976), with reference to work on inclined polystyrene surfaces. As summarized by Collins and Jensen (1983), "... plants are grown in holes in panels of expanded polystrene or other material, but the plant roots are merely suspended in midair beneath the panel, enclosed in a spraying box. The box is sealed so that the roots are in darkness (to inhibit algal growth) and in saturation humidity. A misting system sprays the nutrient solution over the roots periodically. The system is normally turned on for only a few seconds in every 2 to 3 min. This is sufficient to keep the roots moist and the nutrient solution aerated". Additional concepts were developed subsequently by Jensen in Arizona for lettuce.

There is currently no known large-scale commercial applications of aeroponics. Equipment was, however, marketed for plant propagation by this system (in the USA., Israel, The Netherlands). The roots of cuttings are enclosed in a misting chamber and controlled periodically.

4.7.5 Aeroponics II

This culture method was developed in Italy. The main element of an aeroponic plant is the plant carrier tubes, a plastic container in which at set intervals the small plants are fitted, according to the culture. They are placed one above the other on two or more levels, each one parallel with the other. On each level two or more tubes can be found.

The plastic tubes are supported by an iron structure, equipped with mechanical means of enabling the tubes containing the plants to be moved along the lines.

The standard measure of a "plant carrier tube" is 4.50 m in length. The module is therefore a multiple of 4.50 m. The height of the units is 2.15 m. The units are placed in a special arrangement in the glasshouse.

The root part of the plant is placed into the plastic tubes, and a feeding solution is introduced into these containers from time to time. The moist air remaining in the tube between irrigation periods is sufficient for regular plant growth.

The containers, arranged horizontally, may be removed or inserted allowing cultivation to be conducted at several levels and giving high yield in a limited space.

Container placement one over the other follows a precise plan of cultivation. The young plants, at the early stage of growth, are placed in the uppermost containers, to avoid overshading by older, leafier plants; by means of a system of pulleys, these containers may gradually be lowered in order to allow others contact with light, and so the cycle goes on.

Theoretically, almost everything may be cultivated by means of aeroponics. The best commercial yields are, however, from plants with a short cycle, which allows exploitation of the most important characteristics of these techniques.

Outbreak of root disease is limited, and nematodes, etc. cannot develop in this environment.

Cultivation operations are simple and follow a daily timetable which is repeated identically during the whole year. Part of the irrigation work can be handled automatically.

The economics of the system has not yet been proven.

4.7.6 Aeroponics III

Root Mist Technique. In an unusual application of closed-system hydroponics, plants are grown in holes in panels of expanded polystyrene or

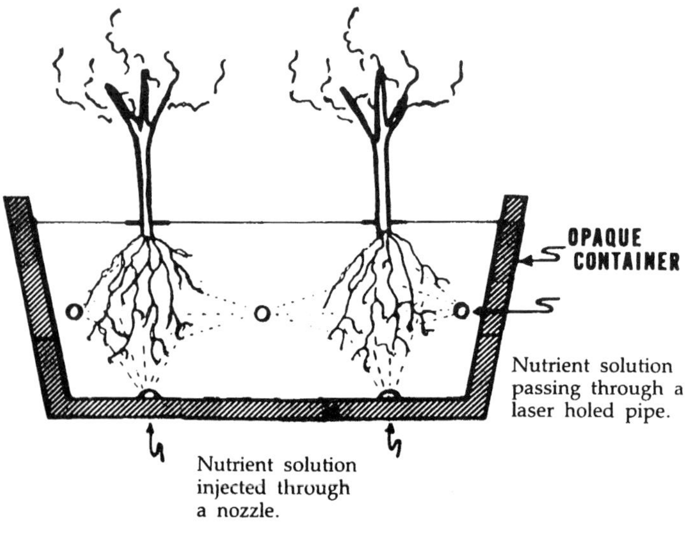

Fig. 4.18. Aeroponic system of growing plants hydroponically showing two different methods of injecting the nutrient mist (Benton Jones 1983)

other material with the plant roots suspended in midair beneath the panel and enclosed in a spraying box (Fig. 4.18). The box is sealed so that the roots are in darkness (to inhibit algal growth) and in saturation humidity. A misting system sprays the nutrient solution over the roots periodically; the system is normally turned on for only a few seconds every 2–3 min. This is sufficient to keep the roots moist and the nutrient solution aerated (Nir 1980).

For some types of advanced horticulture, such as rooted cuttings of imported foliage plant materials without soil, aeroponics may prove to be suitable.

For most culture plants, development of the relatively huge amount of roots is a burden to their energy (carbon) balance toward production of the yield parts of the plant.

There are no known large-scale commercial aeroponic operations, although several small companies market systems for home use.

4.8 Bag Culture – Open Systems

In bag culture the growing medium is placed in plastic bags, which are formed in lines on the (greenhouse) floor, thus avoiding the costs of troughs or trenches and complex drainage systems. The bags may be used for at least 2 years and are much easier and less costly to steam sterilize than bare soil (Fig. 4.19).

Typically, bags are of UV-resistant polythylene, which will last for 2 years; they have a black interior. The exteriors of the bag should be white in deserts and other regions of high light levels in order to reflect radiation and inhibit heating the growing medium; conversely, a darker exterior color is preferable in northern, low-light latitudes to absorb winter heat. Bags used for horizontal applications (the most common) are usually 50–70 l in capacity. When used horizontally, bags are sealed at both ends after being filled with the medium.

The bags are placed flat on the greenhouse floor at normal row spacing for tomatoes or other vegetables, although it would be beneficial to first cover the entire floor with polythylene film. Paired rows of bags are usually placed flat, 1.5 m apart (from center to center) and with some separation between bags. Holes are made in the upper surface of each bag for the introduction of transplants, and two small slits are made low on each side for drainage or leaching. Some moisture is introduced into each bag before planting. Less commonly, the bags are placed vertically with open tops for single-plant growing; these have the disadvantages of being less convenient to transport, requiring more water, and maintaining less even levels of moisture.

Drip irrigation of the nutrient mix is recommended with a capillary tube leading from the main supply line to each plant. Plants growing in

Fig. 4.19. Young tomato plants in bags containing a proprietary peat-based substrate. An irrigation pipe is laid along the middle of the double rows, and the small-bore tubing will later be moved into position to water each plant. The surface of the ground is covered with polythene sheet. (Photograph by courtesy of the Phytology Department, Laval University, Quebec) (FAO 1990)

high-light, high-temperature conditions will require up to 2 l of nutrient solution per day.

The most commonly grown crops in bag culture are tomatoes and cucumbers (also cut flowers). When tomatoes are grown, each bag is used for two crops per year for at least 2 years.

The use of 10- or 12-l plastic buckets is quite common. The buckets are more expensive, but easy to handle. Holes of 2 cm diameter are made in the wide wall, near the bottom, for drainage purposes.

4.8.1 Growth Media in Bag Culture

4.8.1.1 Sawdust and Sand Combined in One Mixture

This has the advantage of permitting a more uniform distribution of moisture than does sawdust alone. This advantage, however, must be offset against the extra cost of supplying and mixing the two ingredients. Not all sawdust is equally suitable (Adamson and Maas 1976).

Although cucumbers can be grown satisfactorily in beds of sawdust, the use of an open-topped plastic bag for each plant allows greater flexibility in sequence cropping (Adamson and Maas 1981). Whereas 10 l of sawdust per plant suffice for tomatoes, 30 l are more satisfactory for cucumbers in order to reduce moisture fluctuations.

Sawdust is also widely used in South Africa for bag culture of tomatoes, cucumbers, lettuce, and other crops (Maree 1981a, Bauerle 1984). After trials with many other soilless media, including straw bales, peat, gravel, and composted bark, sawdust proved not only effective but particularly inexpensive and readily available. The material, from *Pinus pinaster* and *P. radiata*, is used fresh, mixed with wood shavings and without composting.

Sawdust from timber treated with preservatives should not be used. The material is discarded after two crops, thus avoiding the need for sterilization. The bags used are of black polythene and hold 14 l of sawdust. A 2 cm layer of sterilized sand, placed on top of the sawdust immediately after planting, helps to ensure good moisture distribution. Four drainage holes are punched in the bag 2.5 cm above the base.

The recommended conductivity levels for solutions sampled in the bottoms of the bags are 1.6 mS/cm in summer and up to 2.0 mS/cm in winter, with a pH of 6.5.

4.8.1.2 Wood Bark

Good results have been obtained with fresh, uncomposted pine bark chips as a substrate for cucumbers and tomatoes in upright bags. The layout used for tomatoes in open-topped bags, each holding 14 l of bark, has been described by Harris and Maree (1984). Staggered double rows of the bags are arranged with their bases 1 cm below path level, with a drainage channel 5 cm deep along the middle of the row. The entire floor of the plastic tunnel is covered with black 100-μ PVC sheeting to exclude soilborne diseases.

The fresh softwood bark used consisted of pieces 1 cm thick, 2–3 cm long, and 2 cm wide. The plants were grown in 14 l bags, the cultural details being similar to those for sawdust.

Plants growing in woodbark were somewhat retarded during the first 3 weeks, but then caught up and surpassed those grown in sawdust.

The causes considered included low nitrogen and/or phosphorus status, the presence of tannins and plant water stress due to slow wetting (Cotter 1979).

4.8.1.3 Vermiculite

Harris (1976) described a drip-culture system for commercial tomato production in vermiculite. The material was contained in black polythene bags 18 cm × 15 cm and 35 cm in length, laid out in double rows 33 cm apart on a floor covering of polythene. Each bag contained 9 l of vermiculite, and had small drainage holes in the base. The bags were irrigated via a system of plastic pipes and small-bore ("spaghetti") tubing.

Maree (1981b) has described the commercial production of tomatoes in a vermiculite substrate. The plants are placed in 9-l black bags with drainage holes punched not lower than 2.5 cm from the base. The tomato plants are arranged in double rows, 60 cm apart with 40 cm spacing along the staggered rows and paths 1 m wide. A small drainage trench is provided along the center of each double row. Practical points emphasized by Maree (1981b) include the provision of in-line water filters and a simple technique for keeping the ends of the small-bore irrigation tubes within the tops of the bags; the fine tubes are pressed into 2 cm lengths of larger (15 mm) PVC piping. Watering is required at least twice per day, and up to four times per day in midsummer. The nutrient solution must supply all essential major and minor nutrients, since the substrate is relatively inert. At least 10% water above the estimated water requirement should always be provided, the excess running to waste. At the end of cropping, the main roots are cut out and the bags topped up with fresh vermiculite. Even when no crop is being grown, the substrate should be kept moist, since it tends to revert to its original mineral state if allowed to dry out. After being used for two to four crops, the bags should be replaced with fresh material.

The commercial use of vermiculite is questionable.

4.8.1.4 Synthetic Organic Substrates

Several synthetic organic materials have been developed for use as horticultural substrates, including expanded polystyrene, urea-formaldehyde and polyurethane foams.

Amongst the various products available is granulated phenolic-based foam for bag culture, being used commercially for production of tomatoes, cucumbers, peppers, and other greenhouse crops. The material has a stable structure and high porosity (97–98%), water-holding capacity is high and the aeration good. The polythene bags used to contain this material are 15 cm in diameter and either 80 or 120 cm long, giving volumes of 14 and 20 l. The bags are first laid out at the required spacing, after which planting holes are cut in the upper surface and the bags allowed to air for 24 h. Since the material is acidic, the bags are then filled to capacity with a solution of potassium bicarbonate containing 1 g/l. After 24 h, horizontal drainage slits (two to three per bag) are cut at a height of 2 cm from the base. After draining, the bags are then ready for planting. Water and fertilizers are supplied through a conventional drip-irrigation system with an outlet for each plant. Electrical conductivity and pH can be monitored. The recommended pH is 5.5 (5.0–6.5), with conductivities of 2 mS (1.5–2.5 mS) for cucumbers and 2.5 mS (2.0–3.0 mS) for tomatoes and peppers.

The plants are propagated either in bottomless or open lattice pots containing a soilless substrate (Fig. 4.20).

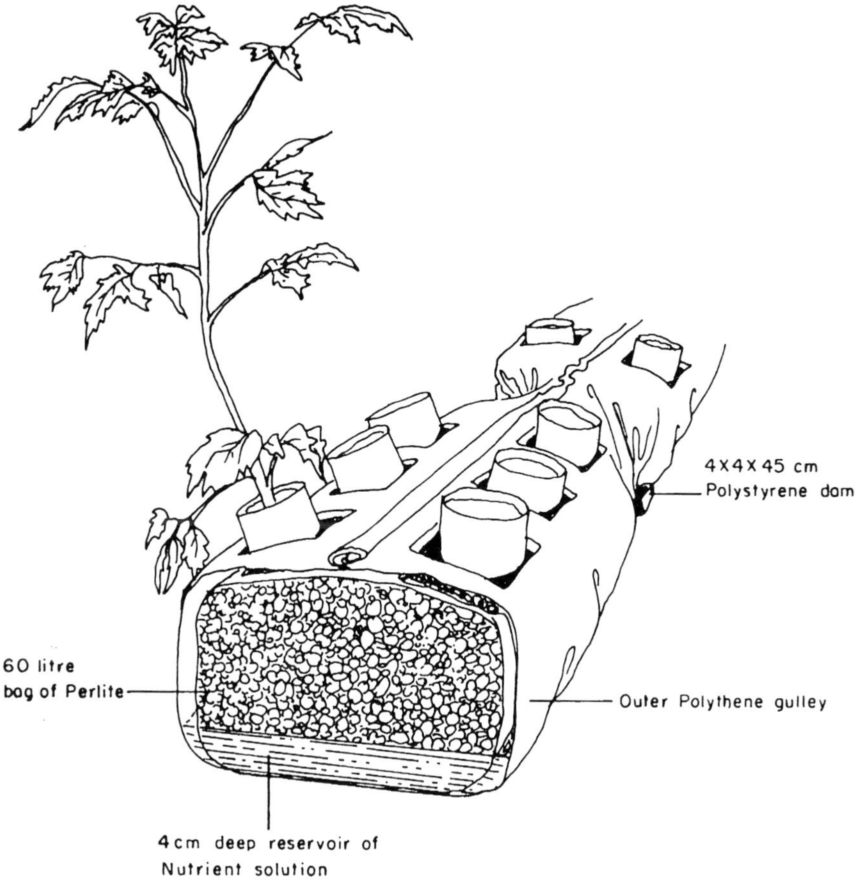

Fig. 4.20. Subirrigated bags of perlite for tomato production. Each bag holds 60 C of substrate and accommodates six plants (Wilson et al. 1984; FAO 1990)

4.8.1.5 Perlite

The use of perlite for hydroponic crop production has been described by Wilson et al. (1984). Tomato plants were grown in sealed bags, laid on their sides, and with holes cut in them to accommodate the plants. Drainage slits were cut in the sides of the bags about 1 inch above the base. The highest yields were obtained when gypsum (10 g/l) and trace elements (0.5 g/l) were incorporated. Wilson et al. (1984) described a system for growing tomatoes in larger (60 l) bags, each bag designed to hold six plants. Instead of being watercd from the top, the bags were laid in lines in polythene gullies. The gullies had lengths of 4 × 4 cm expanded polystyrene placed transversely underneath between the bags, thus creating a series of separate subirrigation regions, 4 cm deep. Vertical slits were made in the polythene walls of the perlite bags in order to ensure rapid movement of water between bag and reservoir (Hall et al. 1988). The commercial use of perlite is still questionable.

4.8.1.6 Peat

Peat bags have been used very extensively in the UK for tomato and cucumber production.

The typical bag for tomatoes holds some 42 l of peat and, laid on its side, supports three plants, though two-plant bags are also used. The paths between the double rows of bags should be raised about 5 cm above the base of the bags, and a shallow drainage channel is provided between the individual lines of bags. Each plant has its own irrigation nozzle, and, once the substrate has been fully wetted, the bags are slit (three slits, 5 cm long, along one side) for drainage.

4.8.1.7 Peat Mixes

Peat mixes for bag culture include peat:vermiculite (50:50), peat:vermiculite:perlite (60:20:20), peat:local sand or gravel (50:50), etc. Often a wetting agent is added.

Some growers add some of the nutrients to the peat mixture as fertilizers. In 1 m^2 of material, the quantities added are approximately: potassium nitrate 1 kg, superphosphate (20% P_2O_5) 1 kg, limestone 3–6 kg, iron chelate (10% Fe) 40 g, and boric acid 30 g. The other nutrients are added to the irrigation water. The pH should be 5.0–5.7 and the amount of the fertilizers added – especially the limestone – should be calculated accordingly.

The best means towards nutrient control is to irrigate the peat mixed with a full nutrient solution, as in rockwool culture, without adding fertilizers to the mixtures.

Two irrigation and drainage methods are common in bag culture using peat or peat mixtures and similar growth media:

1. Full drainage – the slits for drainage are made at the lower part of the bags to allow full drainage. A regular nutrient solution is used as in rockwool culture.
2. Part drainage – the slits for drainage are made above 4–5 cm from the bottom of the bags, so a 4 cm-deep reservoir or nutrient solution remains in the bag. This nutrient solution is soaked up by the growth media, after the plants have used part of the nutrient solution of the growth media (see Fig. 4.21). The advantage is that less frequent irrigation is required, and when part of the nutrients is in the peat mixture, the irrigation procedure is simple and no nutrition mixes are required (Solution A and B) in the irrigation system. The

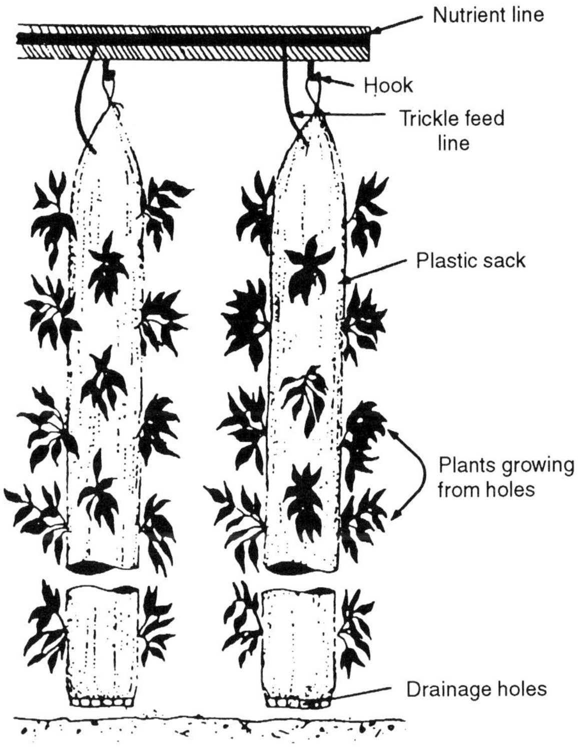

Fig. 4.21. Vertical hydroponics. Growing plants in hanging plastic bags containing perlite or similar material being supplied with nutrient solution (Resh 1983)

nutrient solution reservoir in the bag partially balances the nutrient requirement for the plant.

The disadvantages are: a lower root aeration capacity; a less balanced nutrient solution is available for the plant; and a possible accumulation of salts.

Waterlogging is easy to detect as it results in slow growth and thin plant heads. Waterlogging problems usually develop when the watering regime does not allow enough time for proper aeration between waterings. To avoid waterlogging, make regular frequent checks to control water frequency.

The problem of excessive dryness is equally serious but easy to avoid, provided the irrigation system works effectively and sufficient time is allocated to manage the watering program. If the medium is allowed to dry back frequently (when water cannot simply be squeezed out by hand) plant growth will be impaired, especially if the salt content of the medium is high. Symptoms are a general stunting of growth and considerable yield loss.

The testing of the nutrient solution drainage will advise the grower which amounts of what elements need to be added to the irrigation water, and what is needed to control the pH.

The growth in peat mixes provides less control and balance of the nutrients. The excellent buffering effect of the peat compromises this disadvantage, and economic soilless culture farms use these techniques.

4.9 Other Systems

4.9.1 Vertical Culture

The disposition of the plants is one above the other. This allows for good utilization of solar energy, and a good use of the volume of the greenhouse.

The system consists of a series of vertical "plant-holding" elements filled with an appropriate substrate and supplied with a dripping nutrient solution (Fig. 4.21).

The standard "plant-holding" element that is used as a normal tank, is made of nontransparent plastic about 0.15 mm thick and shaped as a long narrow sack 2 m high and 150 mm in diameter. In its lower part, there are appropriate outlets for the drainage of the nutrient solution which is no longer recycled.

Different materials (i.e., wood-shavings, peat mix, coarse sand, etc.) have been used as a substrate. These materials, both for their physical and chemical characteristics and for their low costs, may be profitable. These substrates have a high water-holding capacity and allow good root aeration. The above-mentioned materials, moreover, limitedly to each vegetative cycle, have proved to be remarkably resistant to degradation.

At the beginning of the vegetative cycle, the "plant-holding" elements are hung in rows on the carrying structures of the greenhouse. Small openings are made in each "plant-holding" container, along several lines and at a reasonable distance one from the other. Through the openings, the plant roots are laid into the sack, leaving the aerial parts outside it. In this way, the roots develop and spread, in the dark hollow of the "plant-holding" element, among the interspaces of the substrate which is periodically supplied with an appropriate nutrient solution.

According to environmental conditions and plant requirements, a dripping nutrient solution is let in from the higher part of the "plant-holding" elements by means of small distribution pipes in such a quantity as to allow a light percolation. The nutrient solution is pumped from a tank one to three times a day for 2 to 5 min, or in intervals by a drip system to the one used in sand culture systems. The waste nutrient is not recycled.

In a new version, square plastic boxes – containers of 20 $\times$ 20 $\times$ 20 cm – were mounted in a diagonal manner, such that each container has an open space where seedlings are planted (Fig. 4.22).

Fig. 4.22. Kohlrabi plants in vertical hydroponics, Kfar, Maimon, Israel

The nutrient solution flows from the upper container to the next lower one through holes in the bottom of the container. The investment in the container is relatively high. The amount of plants per m^2 glasshouse is about 30% higher than in the conventional horizontal arrangement.

This system can be used for smaller plants, such as strawberries, kohlrabi, and pot flowers.

4.9.2 Ruthner System

This system was developed in Austria. Two vertical conveyor belts move within a prefabricated, artificially lighted steel structure that is insulated and lined with reflective material. Plants are moved up and down within a three-dimensional light grid and supplied with nutrient solution at the bottom of the unit.

With systems based totally on artificial light sources, it is difficult, despite marketing claims, to imagine a competitive advantage for such an energy-intensive technology in the everyday world.

5 Water

Crop yields are more dependent on an adequate supply of water than on any other single factor.

5.1 General

The great heat capacity of water causes it to be a temperature buffer – relatively large quantities of heat must be gained or lost by a given quantity of water to effect a given change in temperature. Plants are therefore protected against violent changes in temperature which might upset their steady state. The high latent heat of vaporization of water affects temperature regulation, meaning that a high input of heat is required for the evaporation of a given amount of water (500–600 calories per gram). This property has two important advantages for plants: it minimizes the loss of water, and it provides maximal cooling per unit water lost, since the heat which goes into the evaporation of the water is withdrawn from the plant.

The extremely high surface tensions maintain discrete boundaries where moist cells abut on air and where aqueous phases border on lipid membranes. This same property also promotes the retention of films of water on the growth media particles, when much of the water filling the space among the particles after irrigation has drained away.

Another important feature of water is its cohesiveness, which influences the water movement through the plant body. Water is the best solvent known. Most substances of biological significance are extremely soluble in it. In the context of mineral plant nutrition, the high solubility of inorganic salts in water is especially significant. The growth of plants depends on a constant traffic of mineral elements into and through the plant body, and water serves the function of a carrier of these mineral nutrients. The low viscosity of water permits rapid diffusion of solutes, including inorganic ions through the narrow conduits of the vascular system.

5.2 The Process of Osmosis

If on one side of a membrane there is pure water or a solution, and on the other side another solution but of different concentration, osmosis will occur. The more concentrated solution will have the lower (more negative) water potential. Water will diffuse from the other solution into this one until pressure has been built up.

Several factors in the plant influence this water potential and osmosis. Because in the plant there are many restrictions upon the free diffusion of solutes, the special diffusional water transfer called osmosis is of particular importance. There are cellular membranes, vacuolar membranes, nuclear membranes, membranes around plastids such as chloroplasts, and even membranes around the mitochondria. In each of these examples, water molecules may pass through the membranes more rapidly than solute particles, providing the essential features of an osmotic system. The consequences include not only the development of pressures (turgor) within the cells, but also the bulk movement of solutions within the plant.

The osmoticum can be expressed as osmotic pressure or as osmotic potential. The *osmotic pressure* of a solution is equivalent to that real pressure that may be built up. The *osmotic potential* is that component of a water potential resulting from the presence of solute particles. It is equivalent to osmotic pressure in concept but opposite in sign. It may also be expressed in energy terms, while osmotic pressure is always expressed in pressure terms.

5.3 Transpiration

The control of the water movement from the root environment through the plant to the atmosphere is one of the features of plant growth.

The main absorption or intake of water into the plant is through the roots. The root environment for water uptake is more optimal in most soilless culture conditions than in soil conditions, as discussed in the section on roots.

Absorption of water occurs along gradients of decreasing water potential from the root medium to the root xylem. This gradient is produced differently in slowly and rapidly transpiring plants. This results in two absorption mechanisms: active absorption, or osmotic absorption in slowly transpiring plants, and passive absorption in rapidly transpiring plants where water is pulled in by the decreased pressure or tension produced in the xylem sap (Kramer 1983). This tension in the xylem sap is also dependent on the transpiration rate.

The intake of water is dependent on the temperature of the root environment, dissolved air-oxygen available in the water-nutrient solution, pH, and, of course, on osmoticum (nutrient concentration), plant activity (including diurnal activity), plant age, and climatic factors in the shoot environment.

If the plant is not transpiring, water may be moving only very slowly in the xylem and in the general circulation system. Under conditions of high transpiration water may move very rapidly upward through the xylem. Pressure potential will be the principal component of water potential in the xylem, but since xylem sap contains small quantities of dissolved minerals, osmotic or solute potential will also contribute to water potential.

The resistance of water within the plant will be for the most part relatively stable, but stomatal resistance may vary over a wide range, often in response to changing environment.

5.3.1 Transpiration Rate

Transpiration is the loss of water from plants in the form of vapor. It is basically an evaporation process. It is affected by leaf structure and stomatal opening, and by environmental factors that affect evaporation and stomatal aperture. Transpiration is the dominant process in plant water relations because it produces the energy gradient that causes water movement into and through plants. Excessive transpiration produces midday water deficits and temporary wilting, as it may cause absorption to lag behind transpiration. Permanent wilting may occur.

The rate of transpiration depends on the supply of energy available to evaporate water, the difference in water vapor concentration or vapor pressure between leaves and air, and the resistance in the water pathway toward the leaf surface. Near the leaf surface, the main resistances are the stomata opening, but also the leaf cuticle and the boundary layer surrounding the leaves.

The rate of transpiration is controlled primarily by stomatal resistance, which depends on the degree of stomatal opening. Stomatal opening is affected by internal and external CO_2 concentration (high air CO_2 concentration closes stomata), light intensity, atmospheric humidity, and temperature. Increasing light intensity usually causes stomatal opening; decreasing light causes closure. The interaction between light intensity and CO_2 concentration with respect to control of stomatal opening and closing is complex and varies among species. Stomatal closure can also be caused by dehydration of leaves. The responsiveness of stomata to light, CO_2, and water stress varies with leaf age and is partly controlled by phytohormones such as abscisic acid and cytokinins.

5.3.2 Transpiration Rate Control

As previously mentioned, the transpiration rate is controlled by (1) leaf area and structure, (2) stomatal opening, and (3) those factors which affect the steepness of the vapor pressure gradient from plant to air. Absorption, on the other hand, is controlled by (1) rate of water loss, (2) extent and efficiency of root systems, and (3) the water potential and the osmoticum conductivity of the nutrient solution.

A change in rate of either water loss or water absorption will be instantly transmitted to the other process. However, there is considerable resistance to movement of water through the roots. There is also a buffer in the system in the form of the parenchyma tissue, which functions as a reservoir, losing water during periods when transpiration exceeds absorption, and gaining water when the reverse occurs. As a result, the first effect of a high rate of transpiration is decrease in water content and loss of turgidity of leaf cells, which often culminates in wilting.

The two principal factors influencing the transpiration are the evaporation of water and the stomatal aperture. Increasing leaf temperature promotes evaporation, but may eventually cause stomates to close. Light causes stomates to open, but increases temperature. Increasing humidity decreases transpiration, but relative humidity is a function of temperature, and this in turn is influenced by light. Wind blows away the vapor, causing an increase in evaporation, but if the leaf is warmed by sunlight, wind will lower its temperature, causing a decrease in transpiration.

Wilting occurs when water loss by transpiration is higher than water uptake by the root, or the transport efficiency in the plants is lower than the leaf requirement.

5.3.3 Effects of Water Stress on Plant Growth

Water stress affects practically every aspect of plant growth, modifying the anatomy, morphology, physiology, and biochemistry. Some of the effects are related to decrease in turgor, some to the decrease in water potential and osmotic potentials. In some critical stages of plant growth it causes more injury than at other stages. Especially critical stages are pollination and fertilization.

Water stress also causes a general reduction in size. Characteristic visual symptoms are decrease in leaf area and thicker leaves. There is a decrease in cell size, and an increase in stomatal frequency and cutinization. It also has multiple effects on plant growth. Photosynthesis is reduced by closure of stomata, decreasing the uptake of carbon dioxide. Reduction of energy available through reduced photosynthesis may reduce translocation, accumulation of end product, and translocation of growth regulators.

Temporary, midday water deficits often occur on hot, sunny days because water loss temporarily exceeds absorption. Some plant diseases, such as blossom-end rot in tomatoes, occur more frequently under water stress conditions.

5.3.4 Beneficial Effects of Water Stress

Under some circumstances, moderate water stress can improve the quality of plant products; even vegetative growth is reduced. The taste and quality of some fruits are improved; the aromatic content, nicotine content, and alkaloid content of many plants increase under water stress and salt stress.

5.4 Irrigation

Irrigation should take place while the growth-media water potential is capable of supplying water fast enough to meet the local atmospheric and plant demands without placing the plants under stress that would reduce the yield or quality of the harvested crop. The only soilless culture in which soil water measurements apply is sand culture. Tensiometer measurements may be used.

Although measurement of the plant water content provides the best indications, it is not practical because it involves destruction of plant parts. Measurement of evaporation, whereby the farmer can estimate the water needed by the crop, taking into account all the environmental and plant requirement factors, remains the only viable method of water measurement.

Under soilless conditions, the root system is often reduced in comparison to growth under soil conditions. The water deficit in plants grown under soilless conditions may therefore appear much sooner.

Plants themselves are the best indicators of the need for irrigation. Changes in leaf angle, rolling, or other movements or leaves indicate the development of water stress prior to wilting. This is an indication of overdue irrigation.

5.5 Saline Water

Water is considered to be saline (brackish or salty) and generally unsuitable for practical agricultural purposes if it contains more than 1500 ppm of salts. Specific symptoms of salt excess (toxicity) occur in some plants, while others show only a decrease in yield.

Reduced plant growth under saline conditions has been considered to result from various factors, the most important of which are: "Physiological drought", induced by the low water potential of the soil solution; low water potentials and low "relative turgidity" in plants; osmotic adjustment in the plants as a result of increased ionic concentration in their cells, which may result in deformation of macromolecules by disrupting their shell of bound water, etc. All these changes may result in hormonal imbalance, reduced stomatal aperture, lowered rate of CO_2 uptake; loss of transpiration, chlorosis, and stunted growth – phenomena which are commonly observed in plants grown under saline conditions.

The water balance in plants is a result of the relative rates of movement of water into and out of plants; this movement is a function of both water potential and pathway resistance (which is changed under saline conditions).

Varying degrees and types of damage shown by different plant species subjected to saline conditions are considered to result from the species of ion present in the osmoticum (the overall osmotic potential); the varying ability of roots to exclude toxic salts; the manner of osmotic adjustment; the uptake and concentration within the cytoplasm of toxic ions, and the relative sensitivity of the enzymes to the modified cell environment.

It is difficult to ascribe all of these phenomena as resulting from a special effect, especially since contradictory results are reported by different workers. This may be partly explained by the fact that these types of experiments are conducted under various ionic compositions (although the total salt concentration was similar) and differing root growth media (water, gravel, sand, soil culture methods), which also results in different root aeration conditions and differing root and aerial environmental conditions.

The undesirable effects of an excess of salt in nutrient solutions may, therefore, be divided into four categories: (1) a general effect caused by raising the osmotic pressure of the solution; (2) toxic effects caused by individual ions when present in too large amounts; (3) toxic effects caused by individual ions when present in even relatively small amounts; (4) interference with uptake and, hence, function of the elements caused by ionic imbalance.

5.5.1 Adjusting Saline Water

The use of saline water may be beneficial, since it may contain a sufficient or even significant amount of necessary nutrient elements; however, even if these are present, they usually do not occur in the correct ratio and the water has to be adjusted, a procedure which consists of adding the amount of nutrient elements that are lacking or insufficient in order to correct the ratios between the various constituents.

Since saline water may be adjusted, it may be used in hydroponics, despite the fact that it was once considered unsuitable for agriculture. The adjusting of irrigation water for normal agriculture has shown promise under certain conditions, but it is generally considered impracticable, as is the case in horticulture. The composition of saline spring water used for commercial hydroponics in Israel is given in Table 5.1.

Table 5.1. Composition of saline spring water used in hydroponics in the Arava (Israel)

Salts	Content (in mg/l)
Chlorides	700–900
Sulfates	600–1000
Sodium	400–500
Calcium	350–450
Magnesium	120–180
Total salts	2500–3500
Osmotic pressure	1.5–1.9 atm

Table 5.2. Nutrients added to saline spring water in the Arava (Israel) in closed culture conditions

Nutrients	Amount added (in g/1000 l)
Potassim nitrate	900–1000
Triple superphosphate	200–300
Ammonium sulfate	60
Ferrous sulfate	8–10
Boric acid	1.5
Manganese sulfate	1.5
Zinc sulfate	0.5
Copper sulfate	0.5

The preparation of this water illustrates the principle of adjusting saline water for hydroponic use. The water contains almost twice the amount of calcium and magnesium, and more than twice the amount of sulfate usually added to fresh water. More than 40 times the amount of chlorides, of which only a few milligrams are normally required, are present. The content of sodium, which is not a nutrient, is high. Attempts to precipitate the excesses of these elements were not successful. The water was adjusted by adding extra nitrogen, phosphorus, and potassium; calcium, magnesium, and sulfate salts were not used. The composition of the nutrient solution made up with this water is given in Table 5.2.

5.5.2 Plants Grown with Saline Water

The plants that can be grown under saline conditions are limited to salt-tolerant and moderately salt-tolerant species, such as carnation, tomatoe, cucumber, and lettuce (Meiri and Shalhevet 1973; Meiri and Plaut 1985).

Crop salt tolerance is usually expressed as the expected decrease in yield for a given level of soluble salts in the root medium, as compared with yields obtained under nonsaline conditions (Maas 1986).

Several crop tolerance lists have been published, such as those of the US Salinity Laboratory, the American Society of Agronomy and American Society of Civil Engineers; a suitable abstracted list is given in the Appendix Table 12.

Although the effects of salinity on plants occur physiologically in a relatively low salinity medium, injury symptoms occur under relatively extreme salination. Salt-affected plants usually have a normal appearance, although their growth is stunted and their leaves may not be a darker green, but are thicker and more succulent.

In more extreme cases, especially in woody species, leaf burn, necrosis, and defoliation occur.

Nutritional imbalance is often the first, basic disturbance, and produces specific nutrient-deficiency symptoms.

General stunting of plant growth is the most common salinity effect. With increasing salt concentration the growth rate decreases progressively. Top growth is usually suppressed more than root growth, while the leaf:stem ratio is increased.

Depressed vegetative growth due to salinity does not always relatively affect the commercial crop yield. Grain yields may be reduced without affecting the fresh weight of the plant, as in rice and corn, while in some salt-tolerant plants, as for example, wheat, seed production is decreased less than its vegetative growth. In root crops the most influential decrease is in the storage-root yield, and not in the aboveground parts.

5.5.3 Factors Influencing Salt Tolerance

These include the following: (1) stage of growth; (2) varieties; (3) osmotic pressure; (4) ion uptake, nutrition, and root aeration; (5) environmental factors.

5.5.3.1 Stage of Growth

Sensitivity to salinity varies in each state of growth. Most cereal crops are more sensitive to salinity during the period of seeding growth than during germination and later stages of development. Tomatoes and cucumbers are more sensitive during germination and early development than at later stages. The response of plants to salinity is directly related to their stage of growth and duration of exposure.

The germination of seeds is usually more difficult under saline than nonsaline conditions. Salinity levels that would prevent germination in soil do not always do so in soilless culture; there is no accumulation of salts around the seed due to evaporation, as there is in soil. In soilless culture, the salt level remains more constant, while in soil culture it is being continually raised. Depending upon temperature and other growth factors, tomatoes and cucumbers generally take approximately 20% longer to germinate under saline than under nonsaline conditions.

Toxicity symptoms develop more slowly in older plants. Younger plants are usually more sensitive to salt than more mature plants due to their more limited capacity for water uptake.

5.5.3.2 Varieties

Varietal differences in sensitivity to salinity are much more pronounced than in usually considered. A relatively wide range of variation to this sensitivity is found in barley, wheat, and legumes, as well as in tomatoes and many other culture crops. No listing of salinity sensitivity based on variety is available, a factor which has been neglected in great part in research on varieties. The development of salt-tolerant varieties can be achieved genetically, but this has so far not been given much attention.

5.5.3.3 Osmotic Pressure

The inhibitory effect of saline solutions on plant growth is generally attributed to the increase in their osmotic pressures (the effect of the total concentration of solute particles in the solution, irrespective of their chemical nature), and to the associated reduction in water uptake. Some investigations have shown that water uptake is relatively unaffected by salinity, but this opinion is not general. Extremely high osmotic pressures (over 10 atm) for short periods are less damaging than long periods of moderately high pressures (4–5 atm). The relationship between salinity and osmotic pressures must be carefully watched when saline waters are used for nutrient solutions. High initial salt concentrations and increments of salt by the addition of water to replace losses through transpiration result in very high osmotic pressures. These disrupt water uptake, so that irrigation must be more frequent. More frequent irrigation, despite the wet appearance and frequent replacement of nutrient solutions, leads to a greater use of water than is ordinarily required for soilless culture.

High osmotic pressures cause specific abnormal growth symptoms, the most marked being a marginal leaf burn and bluing and bleaching of the plants' tissues. The first symptom is a generally stunted growth with smaller and darker green leaves. These symptoms are similar to those of plants affected by drought, and may be due to the reduced uptake of water. However, drought-affected plants eventually show a characteristic wilting, while salt-affected plants do not.

5.5.3.4 *Ion Uptake and Nutrition*

Especially important is the effect of saline conditions on ion uptake, irrespective of any osmotic effect. The nutrient requirements of plants vary widely, so that the effect of salinity on nutrition may differ markedly in each species.

Salinity may inhibit the uptake of certain ions. High sulfate concentrations in nutrient solutions promote the uptake of sodium (leading to sodium toxicity), decrease the uptake of calcium (leading to calcium deficiency, especially in lettuce), and interfere with potassium uptake. High calcium concentrations in nutrient solutions also affect potassium uptake. High total salt contents are thought to affect calcium uptake, leading to "blossom-end-rot" symptoms in tomatoes and peppers, blackheart of celery, and internal browning of lettuce, especially in high sulfate growth media. Saline conditions reduce the availability of certain microelements, especially of iron, so that higher concentrations are required than are generally used.

Some research has examined the influence of N and P on the decrease of salt tolerance of certain culture crops, but contradictory results have been reported as well. Many investigations were conducted to determine the effect of a higher K concentration under saline (especially NaCl) conditions.

Specific toxic effects are caused by high accumulations of certain ions in the tissues of plants; these are described in the section on toxic symptoms (Appendix Table 17). Direct toxicity due to sodium and chloride ions of saline water is common, especially in strawberries, roses, etc. These toxic effects occur below the osmotic levels, restricting yields in these crops. Both ions accumulate in relatively high amounts in plant tissues (0.5% Na and 1–2% CI per dry leaf weight), even when the sodium content is not exceptionally high. The first toxicity symptom is leaf burn.

In addition to chloride and sodium toxicity, boron toxicity is relatively common with some saline waters. Boron is translocated to the leaves, and accumulates in their tips and margins. Injury is recognized by a characteristic pattern on growing leaves and fruits. No characteristic symptoms develop in tomatoes, beans, radishes, or peppers; if boron toxicity is suspected, leaf analysis is recommended.

Bicarbonate toxicity, which is fairly common when saline waters are used in normal agriculture, has not, as yet, been observed in soilless culture (Schwarz 1985).

5.5.3.5 *Environmental Factors*

The effect of climate on salt tolerance is largely due to its influence on the transpiration rate, and hence on the rate of uptake of water. Most crops are less salt-tolerant in hot, dry areas, than in cooler, humid ones.

The transpiration rate in soilless culture, grown plants is generally somewhat higher than in soil-grown plants, which is probably also true for saline conditions (Schwarz 1968).

5.5.4 Yields Under Saline Conditions

Another factor to be considered is the decrease in yields. Depending upon species and variety and on the salinity of nutrient solutions, yields may be lowered by 10–25% under saline conditions. The growing of moderately salt-tolerant plants, such as tomatoes, under saline conditions generally remains economically feasible down to a yield decrease of 20%. In Israel the yield of tomatoes and lettuce, grown with water containing 3000 ppm salts, was lowered by 10–15% and that of cucumbers by 20–25%. There was no decrease in the number of carnations grown with water of that salinity; the stems, though thicker, were only 10% shorter. Since salinity leads to stunted growth and a decrease in leaf, bud, and root development, the number and quality of vegetables, fruits, and flowers are affected and the percentage of top quality products is slightly reduced.

Advantages. Not all of the effects of saline water are detrimental. Cucumbers and tomatoes grown with saline-water solution taste sweeter than those grown with freshwater solutions. The sweetness of fruit depends on the plant's sugar to organic acids ratio, which seems to be greater under saline conditions.

Plants grown in saline solutions apparently have a much higher tolerance to zinc and copper, so that zinc and copper levels previously considered toxic may be present without causing injury (Schwarz 1985).

5.5.5 Carbon Balance in Salinity (Maintenance Repiration)

One of the main factors expected to reduce plant growth under salinity is the reduction of photosynthesis. Moderate salinity affects photosynthesis by reducing stomatal aperture, even when there is full adjustment of the internal osmotic concentration, and turgor is high. This interferes with carbon dioxide diffusion into the plant. The carbon dioxide concentration will be the limiting factor of photosynthesis, whenever light intensity is high and the basic light reactions and biochemical pathways of photosynthesis are not damaged.

Growth (increase in weight) is often reduced at low levels of salinity. This level may vary greatly between plant species. Increasing salinity usually results in a proportional reduction in growth until salinity attains levels where apparent signs of damage are observed, e.g., chlorosis and leaf burn.

The reduction of growth under low salinity could be related to the increase of maintenance respiration. The energy cost of maintenance is greater under salt conditions as a result of requirements for compartmentation and secretion and for repair of cellular damage. More plant assimilate resources are required for maintenance, leaving less for plant growth, which is thus reduced. The energy cost of maintenance respiration in salinity was calculated in some salt-tolerant species to be 30% of the decreased growth.

It has been suggested that the salt-induced increase of respiration during the day is a result of the plant being unable to use assimilates for growth. It was found that CO_2 fertilization stimulated growth more in salinized than in control plants. This supports the hypothesis that at low salinity the plants suffer from a shortage of assimilates due to reduction of photosynthesis and an increase of maintenance respiration.

The carbon fixation efficiency (carbon remaining in the plant after 24 h) (carbon fixed-maintenance respiration) was found to be about 0.75 for all plants and is not affected by salt.

Morphological disorders under low saline conditions, e.g., reduction of leaf thickness, leaf areas, and internode distances, are expected to be amongst the signs of reduced growth. Recent research indicates clearly that these disorders are also a result of reduced available energy in salinized plants for plant structure, and not a result of salinity per se, as most of these disorders could be alleviated by increasing the CO_2 level (Schwarz 1985; Schwarz and Gale 1981, 1983).

5.5.5.1 Multiple Stresses

The major obstacle to increasing the tolerance of crop plants to physical and chemical stresses is the lack of fundamental knowledge of the basic mechanisms of stress injury and stress tolerance, and the plant growth-limiting factors in these conditions.

5.5.5.2 Plant Responses to Stress Factors

The effects of drought, salinity, and heat on plants are usually multiple and often interrelated. Studies of the responses of plants to physicochemical stresses show that many biophysical and biochemical steps in metabolism can be affected. From this complex of multiple responses, it is often difficult to identify the primary site of damage. Thus, many response factors must be studied to determine the threshold of damage for different responses and to understand the relationships that might exist between the primary effect and the cascade of processes that are, in turn, affected.

Drought. Water stresses reduces or arrests plant growth because of a variety of effects. The carbon fixation steps of photosynthesis stop because carbon dioxide exchange into the leaf is blocked as the stomatal pores close to halt further water loss from the plant via transpiration. *Heat* energy from solar radiation is no longer effectively dissipated through the evaporation of water, and leaf temperatures may rise to damaging levels. Solar energy, absorbed by photosynthetic pigments, is no longer channeled to carbon fixation, and the photosynthetic system that can be inactivated as energy is diverted to harmful reactions. Without sufficient turgor pressure, the cells at the growing tips of plants cannot expand and elongate. Growth is also arrested because the products of photosynthesis needed for cell wall protein biosynthesis are in reduced supply. Plant growth substances involved in regulating the opening and closing of tomatoes are likely mediators of plant growth changes brought about by water stress.

The number of possible effects in the cascade of responses to drought will depend on the severity and duration of the water stress. Some responses, such as stomatal closure, are quickly reversed when water is added. The kinds and degree of damage and the rates of repair or replacement are unknown. Also, drought may trigger long-term developmental changes in morphological growth pattern that limit flowering, pollination, and seed development, which can greatly reduce crop yield.

5.6 Wastewater (Sewage)

Sewage effluents are highly enriched with plant nutrients and are usually beneficial to agronomic crops when used for irrigation. A balanced nutrient solution for utilization in soilless culture can be prepared from settled domestic sewage (Schwarz 1973; Berry et al. 1977; Butler et al. 1988).

5.6.1 Treatment with Domestic Wastewater Effluents

Fundamentally, treatment with wastewater consists of the removal of settled solids; suspended solids (mostly organic), and dissolved organic matter measured a Biochemical Oxygen Demand (BOD), and Chemical Oxygen Demand (COD). This treatment is accomplished by physical settling and through bacterial decomposition or digestion of the organic matter either aerobically or anaerobically. The above treatments are designated as primary and secondary treatments. Tertiary treatment is usually required wherever limitations exist in the quality of the effluent discharge into the receiving body of water (sea, lake, reservoir, river, or stream). This would involve, for example, the removal of nutrients favoring algal growth (phosphate and nitrogen compounds, etc.) in order to reduce eutrophication trends in the natural water system. Tertiary treatment is imperative whenever sophisticated waste water reclamation is planned, and includes the reduction of total nitrogen, residual suspended solids, bacteria and viruses, and refractory organic substances usually measured as total organic carbon (TOC). This is usually effected by chemical clarification with lime, and absorption on activated carbon. Sometimes, filtration, chemical neutralization, and chlorination are also added to the process. The residue of the above-mentioned chemicals must be taken into account when these water sources are considered for use in soilless culture.

5.6.2 Time of Sewage Use, According to Fluctuation

The most important fluctuations in sewage are:

1. *Daily fluctuations.* concentrations of ions and impurities are higher during the night than during the day; often important differences may also be observed during the day.
2. *Weekly fluctuations.* on weekends the sewage is more concentrated.
3. *Seasonal fluctuations.* water consumption in winter is reduced as compared to the summer, and thus the sewage will be more concentrated.

The daily fluctuation is less pronounced if there is a considerable distance between the community or town and the sewage collection site so that the sewage collected over various hours will have become mixed during the flow in the pipes and in the settling tanks. If the retention time in the sewage treatment installations requires several days, as practiced in several primary and secondary treatment methods, the daily fluctuation count can be omitted, since the sewage content will be similar on all occasions.

It should be noted that although extraordinarily high sewage concentrations for short periods are not unusual, it should be investigated if the sewage effluent assumes a different appearance or an odor is noticed. This may be due to large amounts of oil or of a chemical material being washed into the sewer. No general or average test of the sewage will include such sudden occurrences so that the farmer must be on the lookout for them.

pH variation: according to type of sewage treatment and fluctuations.

Inorganic nutrient content: the average content of sewage after various treatments is listed below (Table 5.3).

Most of the nitrogen is present as ammonium and usually no further ammonium should be added. The nitrate-nitrogen content is always low, and more nitrate should be added, usually KNO_3. The nitrate-ammonium relationship in sewage is not stable due to bacterial activity and pH fluctuation. There is no general rule as to how often the nutrient solution should be tested to assure the amount and relationship of the nitrogen forms, since the buffer effect differs in various types of domestic sewages. A more frequent testing schedule when a new sewage source

Table 5.3. Composition of treated sewage (averages)

	After primary treatment (settled sewage)	After secondary treatment (activated sludge effluent)
Total solids mg/l	800–1300	200–800
Total suspended solids mg/l	80–120	5–30
Volatile suspended solids mg/l	45–85	5–30
BOD (5 days) mg/l	110–180	10–25
COD mg/l	220–310	20–40
Total hardness (as $CaCO_3$) mg/l	120–200	110–170
Total alkalinity (as $CaCO_3$)mg/l	140–210	70–130
pH	6.8–7.6	7.0–7.8
Total nitrogen mg/l	30–45	20–35
Organic N mg/l	8–25	5–20
NH_3 – N mg/l	14–25	0.5–12
NO_2 – Nmg/l	0–0.5	0–0.3
NO_3 – N mg/l	0.4–3.4	15–30
Total phosphorus mg/l	10–15	8–12
Organic P mg/l	0.3–1.0	–
Inorganic P mg/l	9–13	8–12
Chlorides mg/l	65–150	65–150
Sulfates mg/l	80–120	80–120

According to Standard Methods for the Examination of Water and Waste water, Am Public Health Assoc, 12th edn

is being used will certainly provide the required information to a skilled soilless grower.

The phosphorus content of sewage is relatively low, as compared to usual nutrient solutions. Since 2 mm of phosphorus will often be sufficient for the healthy growth of higher plants, and because this amount is often present in sewage effluent, no further addition of phosphorus is needed except for possibly small amounts. Since sewage effluent is for the most part basic, precipitation of phosphorus, often together with iron or iron compounds, is to be expected; the addition of higher amounts of inorganic phosphorus at a concentration of 5 mm or more may cause iron deficiency effects if the concentration of phosphorus in the nutrient solution is too low, small quantities of phosphorus, up to 1–2 mm, should be added more frequently, probably according to testing schedule, every second day. Phosphoric acid is the form of additional phosphorus most often suggested, since the pH is influenced as well.

Large amounts of phosphorus are attached to organic compounds in detergents, which are released into the water during the deformation and decomposition of sewage organic matter. Washing customs and the types of detergents available determine the quantities of this phosphorus source, and only testing of the nutrient solution will provide the required information of the available phosphorus.

Although the *potassium* level in sewage will depend to some extent on the type of soap used in the community, inorganic potassium generally has to be added and can be combined with the nitrogen supplement as potassium-nitrate.

The *calcium* level available is usually too low in sewage, and thus calcium has to be added (as calcium nitrate or similar forms).

Raising the *magnesium* concentration to normal requirements can best be achieved by adding magnesium sulfate, which provides *sulfur*, a macroelement often too low in concentration in sewage effluent.

Iron deficiency may be suspected, even if determination of iron in sewage tests, are positive, as part of the iron is concentrated in complexes and is not available to plants. According to test information, adding small amounts of iron (2–3 g/1000 l) every second day seems to be a necessity.

The amount of *zinc* (average 13 mg), *copper* (average 1.1 mg), and *boron* (average 1.0 mg) is usually sufficient for higher plants. The *manganese* concentration may be too low (average 0.3 mg) and should be added.

5.6.3 Health Factors in the Reuse of Waste Water

A major factor which must be considered when directly reusing treated waste water for irrigation is the degree of its contamination with pathogenic organisms. Domestic sewage, before it is treated, usually contains the complete range of pathogenic microorganisms found in the community that produced the sewage.

It is generally accepted that the absence of coliform organisms serves as the main indication of the bacteriological efficiency of waste-water treatment processes. In primary sedimentation, a reduction of 40% in the number of coliform organisms is obtained, while in most full biological treatment processes the reduction is between 90 and 98%.

Most countries have health regulations which restrict irrigation with partially treated waste water to specific crops that are not used for direct human consumption, which, of course, does not include the growing of flowers, etc.

6 Carbon Dioxide

Carbon is a plant nutrient. The uptake of carbon by the plant is different from that of inorganic nutrients. The inorganic nutrient uptake is mainly through the roots in the liquid phase, but carbon uptake is through the shoot, in the gas phase, such as carbon dioxide.

Low carbon concentration is the limiting growth element factor under many soilless-culture conditions.

Carbon dioxide is taken up by the plants through stomata, which are primarily positioned in the epidermal layer at the underside of the leaf. Carbon dioxide uptake is primarily a process of diffusion, and the rate is thus dependent on the CO_2 concentration difference between the greenhouse air and the air inside the stomata. Increasing the CO_2 concentration outside the leaf will, therefore, increase the rate of uptake.

Outside air contains about 340 parts per million (ppm) or volume per million (vpm) of CO_2 by volume. Most plants grow well at this level, but if levels are higher, the plants will respond by producing more sugars. During the day, in a closed greenhouse, the plants use the CO_2 in the air and reduce the level below the normal 340 vpm. This is the point at which CO_2 addition is not important. A drop in carbon dioxide levels below ambient has a stronger effect than supplementation above ambient. Actually, the decrease in photosynthesis when carbon dioxide levels drop from 340 to 200 vpm is similar to the increase when CO_2 levels are raised from 340 to about 1300 vpm. In new greenhouses, and especially in double-glazed structures which have reduced exchange rates, the carbon dioxide levels could easily drop below 340 vpm and have a significant negative effect on the crop. Ventilation can bring CO_2 levels closer to ambient but never back to 340 vpm. Supplementation of CO_2 is seen as the only method to overcome this deficiency and, moreover, to increase the level above 340 vpm, which for most crops is beneficial (Turban 1978; Blom et al. 1990).

The benefits of CO_2 enrichment may be evident as larger yields – increase of harvest index ratio, a higher quality product, or shorter time to harvest.

Many growers add CO_2 simply because it is what everyone does, not realizing that CO_2 is actually a plant nutrient, with deficiency and toxicity effects, and the law of the minimum applies to CO_2 just as it does to other nutritional or environmental factors. Only when all the other growth factors, including nutrients, are in proper or near-proper rates will increasing the CO_2 concentration in the air have a beneficial effect (Kramer 1981).

Following is a summary of many reported findings in this area:

Carbon Dioxide Requirements

The actual needs for CO_2 will vary with

1. plant type (C_3 or C_4 plant),
2. leaf area index (LAI),
3. stage of development or crop (vegetative, reproductive, etc.),
4. efficiency of use of light by canopy,
5. nutrition and water supplement,
6. ambient temperatures – day and night/root and shoot,
7. dark respiration and photorespiration rates,
8. photosynthetic efficiency at different light levels,
9. stomata closure.

The net CO_2 requirement is determined by the net rates of photosynthesis. The mass of CO_2 required per mass of carbohydrate is in the ratio of molecular weights of CO_2/CH_2O, which is $44/30 = 1.4$.

Very high rates of photosynthesis measured are 200 t/ha, or approaching 120 g/m^2/day or 13.7 g/m^2/ha.

If the CO_2 is provided entirely by ambient air, the calculation is as follows (assuming the ambient air is allowed to drop from 320 vpm to 220 vpm = removing 100 volumes of CO_2 from every million volumes of fresh air):

1. Density of CO_2 at 30°C is 1780 g/m^3.
2. To supply 20 g/m^2/h of CO_2 requires 0.011 m^3/m^2/h of CO_2 or 110 m^3 air/m^2/h.

This corresponds to an exchange rate of approximately 0.92 volumes min^{-1}, in a 2-m- high enclosure, or a change of air almost every minute.

6.1 General Observations

Young tissues benefit more from CO_2 enrichment than do older tissues in C_3 plants. Enriching C_3 plants with three times the current ambient levels of CO_2 (1000 vpm) raises the average mean yield of immature plants by a ratio of 1.5 (50%), and of mature plants by 1.25 (25%).

In C_4 plants, the vegetative growth is stimulated only by an average of 1.15 (yield ratio), while many mature plants are enhanced by twice the yield ratio.

Clearly, the generalizations relating vegetative growth to reproductive growth are not yet fully supported, although young C_3 plants may appear to have a significant growth advantage at elevated levels of CO_2. C_4 plants respond particularly well to CO_2 at high irradiance, and the increased seed production by a mature C_4 plant may favor the long-term dominance of C_4 plants in mixed open-field populations (Enoch et al. 1972).

Plants receiving CO_2 enrichment may have a different biochemical composition and a different morphological appearance (thicker leaves, more side shoots, etc.) With increasing CO_2 concentrations (up to 1500 vpm), stomatal resistance increases and transpiration decreases on plants grown at low light intensities. High CO_2 concentrations increase salt resistance in tomato plants grown at low light intensities (Schwarz and Gale 1983).

6.1.1 Vegetative Growth

CO_2 enrichment enhances plant growth, especially vegetative growth. Insofar as plant productivity can be related to leaf photosynthesis (i.e., CO_2 fixation), the effect of CO_2 enrichment seems predictable. Although leaf carbon fixation may be two- to threefold greater at 1000 vpm CO_2 than at 300 vpm, this increased carbon fixation by a source leaf is not necessarily matched by an equivalent increase in whole plant dry matter production or yield over an extended growing period.

6.1.2 Leaf Crops

Lettuce yields increase 30–50% when plants are grown in 600 vpm CO_2, but higher concentrations of CO_2 may provide little additional benefit. Lettuce grown in 2000 vpm CO_2 had a greater leaf area than controls, but did not have a greater dry weight under identical light and temperature conditions. What is the limiting factor? Many leaf crops indicate that CO_2 enrichment enhances vegetative growth (1.44 mean yield ratio) to the same extent as mature yield (1.40 mean yield ratio). The question arose, why is no accumulative effect obtained and yet a multiple yield observed?

6.1.3 Root Crops

Normally, CO_2 enrichment of the shoot results in an increase in an root growth. The root is a major sink of new photoassimilates, and CO_2 enrichment results in substantial increases in the yield of those crops in which root tissue represents the marketable organ.

6.1.4 Reproductive Development

Flowers: the development of consumer demand for fresh flowers during the winter season has stimulated improvement of flower-crop management for the greenhouse. Work has focused on high-value floriculture crops such as roses, carnations, and chrysanthemums.

High CO_2 increased the flower production of roses, apparently by reducing the proportion of blind shoots to blossoms. Overall plant size is increased and greater numbers of side shoots and buds have been observed. An increase in flower number can also be obtained with increased light levels, emphasizing that response to CO_2 enrichment is affected by other environmental parameters (e.g., light, temperature, nutrient supply) which may be limiting to plant productivity (Zieslin et al. 1986).

6.1.5 Fruit Development

The most extensively studied horticultural crops in which fruit production has been studied under CO_2 enrichment are tomato and cucumber. While most experiments report only total fruit yield, some yield analysis data have been accumulated.

An increased number of fruit in tomato apparently contributes more to CO_2-induced yield increased than gains in individual fruit weight. Tomatoes grown in high CO_2 produce more trusses per plant, more open flowers per truss, and more fruit per truss. The effects of CO_2 enrichment are more pronounced during the early part of the harvest period, in part because anthesis, and therefore initial yields, are advanced. As a result, the enhancement of tomato fruit production by CO_2 enrichment may appear to be greater during the early part of the season than during later periods.

Total fruit yield of cucumbers grown at high CO_2 concentrations was higher than that of control plants grown at 300 vpm. The yield enhancement during the first month of harvest was 100% higher under the CO_2-enriched condition, which may be attributed to earlier flowering, increased number of flowers, and a shortened harvest period. The increase in fruit number may be further attributed to the increased branching of enriched plants and the higher proportion of female flowers found on the side shoots relative to the main shoot.

Accelerated growth during the relatively immature stages of tomato and cucumber development may prove to be the most significant effect of CO_2 enrichment on yield in these crops in some plants. Greater axillary bud development under high CO_2 leads to shorter flower, and fruit development, which may result in a larger sink demand for photosynthate. In many crops, increased sink demand has been shown to enhance photosynthetic rates in source leaves. Thus, the basis for the beneficial responses to CO_2 enrichment appears to have several components. First, the maintenance of photosynthetic activity and the enhanced rate of photosynthesis favor greater photoassimilate movement and the development of healthy growing sinks. The maintenance of elevated levels of CO_2 during fruit filling ensures an adequate supply of CO_2 for photosynthesis to meet increased demand (Porter and Grodzinski 1985).

6.1.6 Effects of High CO_2 Concentration

Leaf plants:

– Lettuce, 30–50% increase in fresh weight, larger leaf area, crops 4 days earlier, no increased dry weight.

– Tomato, more trusses, more fruit, especially during early parts of harvest.

– Cucumber, up to 100% more fruit during first month of harvest, earlier flowering, more flowers, more female flowers.

– Pepper, greater number of fruit.

Flowers:

– Chrysanthemums, greater number of cuttings in seedlings, longer stems, larger leaf area, larger blooms, increased fresh and dry weight.

– Rose, reduced blind shoots, increase in size and number of flowers.

– Gerbera, no increase in number of flowers, but more flowers of superior grade.

– Cut flowers, in general longer, thicker and stronger stems.

Root crops:

– Potato, more tubers (if CO_2 is applied to roots).

Toxic by-products associated with combustion can affect plant productivity. The effect of SO_2 as a by-product is the reduction of photosynthetic rate, presumably by causing the stomata to close. Sulfur-free fuel should be used in greenhouses.

Nitrous oxide (NO_x) concentration varies diurnally because the conversion of the initial by-product NO to NO_2 is mediated by light and the presence of oxygen. This concentration of either NO or NO_2 reduces tomato leaf photosynthesis by approximately 30%, which counteracts the growth advantage of a 1000 vpm CO_2-enriched atmosphere.

Chronic exposures to a low level pollutant may be injurious. Low level combinations of pollutants such as SO_2 and NO_x may cause greater effects than either pollutant alone at higher levels. This could be especially important when greenhouses are located in industrial areas where airborne pollutants are a constant component of the background.

Carbon monoxide (CO) released as a result of CO_2 combustion does not present a serious problem to plants. Humans are much more sensitive to CO than are plants. Most CO production is due to poorly maintained burning equipment. A small amount of CO is released by plants under photorespiratory conditions (high amount of light, high temperature, and low CO_2 levels) but does not have a deleterious effect, particularly when CO_2 levels are high.

In some burning processes, ethylene, as well as propylene, a C_2H_4 analog, may be released in small amounts. These gases are growth regulations and their accumulation in a greenhouse could have deleterious effects on the plants. Regular ventilation procedures overcome this problem.

Higher concentrations of CO_2 (over 2000 vpm), if supplied over several weeks, can cause senescence. This is expressed as a loss of chlorophyll – earlier leaf yellowing.

Proper burner maintenance and adequate O_2 supply (good ventilation) prevent the accumulation of gaseous pollutants (Blom 1990).

6.2 CO_2 Sources

6.2.1 Combustion

Many growers still favor this method as the raw materials (e.g., natural gas, propane, kerosene, or paraffin) are available, reasonably inexpensive, and produce heat as a useful by-product. The equipment for generating CO_2 by efficient combustion of fossil fuels has been commercially available for many years.

Major problems can arise with combustion. The water vapor generated during combustion can complicate humidity control inside a greenhouse, particularly during winter months when excess water vapor condenses on the cooler outer walls. During warmer seasons, the associated heat production may necessitate earlier venting and loss of CO_2. In addition, potentially toxic pollutants can be generated, particularly if combustion is incomplete due to low flame temperature or restricted O_2 supply.

Pure CO_2. The CO_2-enrichment method with the fewest side effects and problems is injection of pure CO_2 from pressurized tanks of liquid CO_2 directly into the greenhouse circulation system. Commercial growers generally choose between liquid CO_2 and combustion on the basis of the relative cost of local fuel and compressed CO_2, as well as their proximity to a supplier of liquid CO_2. For research purposes, particularly controlled-environment studies, injection of pure CO_2 tends to be used almost exclusively. Dry ice CO_2 is not widely used because of the difficulties associated with its storage, handling, and dispensing.

Nontraditional Sources of CO_2. Numerous industries (e.g., electrical power, cement, and petroleum product producers) exhaust CO_2 as a by-product of their manufacturing processes. Industrial CO_2 represents a massive source of “free” CO_2 that might be used for crop production. The presence of other gases that are toxic to plants may limit direct use of industrial CO_2 in greenhouses until adequate scrubbing techniques are developed.

6.2.2 Monitoring CO_2 and Detection of CO_2

For optimal crop response, CO_2-enrichment systems should be linked both to environmental conditions (light, temperature, water and nutrient application, and ventilation) and to the growth and development of the crop (vegetative, flowering, and fruit development). Presently, commercial and experimental systems tend to link control of supplementary CO_2 with environmental variables such as light or temperature in the greenhouse rather than with the status of the plants themselves. Efficient use of CO_2 may be more effectively achieved by monitoring the CO_2 concentration in the greenhouse, which reflects the photosynthetic activity and CO_2 demand of the crop.

Electrochemical detection is based on the linear increase in the electrical conductivity of water with increasing dissolved CO_2 and HCO_3^- content. A stream of sample air is bubbled through water of known temperature and pH, and variation in the CO_2 content is recorded as a change in solution conductivity. The error in measuring air levels of CO_2 by this method can be as great as 5–10%, even when the temperature dependence of the solubility of CO_2 in water is taken into account.

A more direct method of analyzing CO_2 concentrations is to directly measure the gaseous CO_2 content of the air on a continuous or intermittent basis. Several systems are now sold commercially which allow this approach in greenhouses. At the heart of these systems is an infrared gas analyzer (IRGA), which measures specific absorption by CO_2 in the infrared region of the spectrum.

In most commercial CO_2-monitoring systems, the IRGA operates in the absolute mode and is sensitive over a rather broad range of CO_2 concentrations (e.g., 0–2000 μl/l). In contrast, the differential mode (the reference gas may be any known CO_2 mixture) is preferred for photosynthesis and gas exchange research purposes because greater accuracy can be achieved over a smaller range of CO_2 concentrations.

The CO_2 content of an air sample may also be monitored colorimetrically. A discrete sample or a continuous stream may be bubbled through an indicator solution that changes color in response to CO_2-generated pH changes. Coupled with a photoelectric detection system, the CO_2 concentration may be monitored. Like the conductivity system, this method is relatively slow and temperature-sensitive because equilibrium must be attained between the CO_2/HCO_3^- in the gas stream and the solution.

A simple CO_2-tester consists of a small hand pump which passes air through a tube. The tube contains a CO_2-sensitive chemical which

Table 6.1. Mean yield response of crops to CO_2 enrichment. (After Kimball 1983a, b)

Crop	Immature crops		Mature crops	
	Mean ratio[a]	N[b]	Mean ratio	N
Vegetative				
Cabbage (*Brassica oleracea*)	1.28	10	1.06	2
Clover (*Trifolium* sp.)	–	–	1.12	2
Endive (*Cichorium endivia*)	–	–	1.15	1
Fescue (*Fescus* sp.)	1.51	2	–	–
Lettuce (*Lactuca sativa*)	1.68	7	1.35	54
Swiss chard (*Beta vulgaria* var. cicla)	–	–	1.67	17
All Leafy crops (average)	1.44	19	1.40	76
Root and tuber crops				
Potato (*Solanum tuberosum*)	–	–	1.44	18
Radish (*Raphanus sativus*)	1.79	6	1.28	5
Sugar beet (*Beta saccarifera*)	1.75	7	–	–
All root crops	1.77	13	1.40	23
Flower crops (number of blooms)				
African violets (*Saintpaulia* sp.)	2.36	3	–	
Carnation (*Dianthus caryophyllus*)	–	–	1.07	33
Chrysanthemum (*Chrysanthemum* × *monfolium*)	–	–	1.06	66
Cyclamen (*Cyclamen persicum*)	–	–	1.35	3
Nasturtium (*Tropasolum majus*)	–	–	1.86	3
aGaranrama (*Polargonium* × *hortorum*)	1.19	3	–	
Rose (*Rosa* spp.)	–	–	1.27	24
Snapdragon (*Antirrhinum majus*)	–	–	1.03	1
All flower crops (average)	1.28	6	1.12	130
Fruit crops				
Cucumber (*Cucumis sativus*)	1.46	26	1.43	31
Eggplant (*Solanum melongena*)	–		1.88	3
Grape (*Vitis vinifera*)	2.49	1	1.88	–
Maskmelon (*Cucumis melo*)	–		1.13	7
Okra (*Hibiscus esculentus*)	2.74	3	–	
Pepper (*Capsicum annuum*)	2.41	1	1.60	4
Strawberry (*Fragaria* × *ananassa*)	–		1.17	13
Tomato (*Lycopersicon esculentum*)	1.52	24	1.17	131
All fruit crops (average)	1.56	55	1.22	189
Woody crops				
Apple (*Malus pumila*)	1.32	4	–	–
Birch (*Betula* sp.)	1.06	4	–	–
Cotton wood (*Populus deltoides*)	1.69	2	–	–
Fir (*Abies* sp.)	1.18	2	–	–
Gum (*Nysse* sp.)	1.10	3	–	–
Lees sp.	2.58	1	–	–
Maple (*Acer saccharinum*)	1.74	2	–	–
Pine (*Pinus* sp.)	1.36	26	–	–
Spruce (*Picea pugens*)	1.58	17	–	–
Sycamore (*Platanus occidentalis* L.)	1.21	2	–	–
All woody crops (average)	1.40	63	–	–
All C_3 plants	1.51	211	1.25	542
All C_4 plants	1.11	14	2.01	5
All species	1.48	225	1.26	547

[a]Ratio of CO_2 = enriched yield to nonenriched yield.
[b]Number of observations.

changes color as CO_2 is absorbed. The pump is stroked a given number of times. The length of the tube which changed color is measured on a scale which directly indicates the level of CO_2 in the air passed through the tube. The tubes are disposable (Porter and Grodzinski 1985). Some mean yield responses of crops to CO_2 enrichment are given in Table 6.1.

6.3 Economics of CO_2 Enrichment

It is estimated that about 0.5–0.6 kg of CO_2 per hour per 100 m^2 must be added in a "standard" glass greenhouse to maintain 1300 ppm. For double-polyethylene houses the supplementation should be about 0.25–0.35 kg of CO_2/h/100 m^2.

In general, growers add CO_2 only during the winter months, and an estimate of 1000 h of supplementation per year appears adequate. These values would require 5 to 6 kg of CO_2/m^2/year for a glasshouse and one-half this amount for a plastic house. Table 6.1 presents the current cost of CO_2 addition using the following fuel prices: 15 cents/m^3 of natural gas and 25 cents/l of propane.

The approximate annual cost of fuel for CO_2 supplementation (not including equipment depreciation costs) is natural gas – 45 cents/m^2, propane – 75 cents/m^2.

The price of liquid CO_2 (1 kg is 10–15 cents) would be about 70 cents/m^2.

The depreciation of the equipment adds very little to this cost. In most situations the higher yields justify this expense.

Following are some notes for CO_2 calculation:

1 kg of CO_2 is equivalent to 570 l of CO_2;
1 m^3 of natural gas provides about 1000 l of CO_2 when combusted;
1 m^3 of natural gas = 0.75 l of kerosene = 1.0;
1 of propane for an equivalent amount of CO_2 produced.

CO_2 controllers are used in most greenhouses (see Chap. 9).

Future developments are expected in the breeding and selecting of cultivars adapted for optimal performances in high CO_2 concentrations. These cultivars, that would not close their stomata in high CO_2 will have improved photoassimilate partition, high water use efficiency, etc.

7 Root Environment

7.1 Root Functioning

Root functioning seems to be the bottle-neck in production in most situations. Root functioning is controlled by the interplay between shoots and roots and root respiration rates are influenced accordingly.

Energy supply, or the respiration rate for root functioning, is related mainly to activity of basic maintenance respiration; water and nutrient uptake, and transfer and metabolite transfer and storage. The interplay between too much and too little root functioning can be extremely important in crop plants, especially in saline conditions.

Root development also depends on root media and root environment, e.g., many hydroponically grown plants have no rootlets and shoot-to-root ratios measured in dry matter, root surface area, etc. differ to the extreme in the same plant in various growth conditions. These basic differences may influence yield. The waste of energy for producing plant tissues and organs and the need for more energy for maintenance respiration of this "not important" plant material may be of economic value.

The knowledge of the daily course of the energy used, the respiration rate of the roots in regular and saline conditions, may help to understand the root requirements and open up a possibility for controlling and/or supplying adequate O_2 for this respiration rate (Schwarz 1984).

The deficiency of oxygen or the excess of CO_2 in roots will greatly reduce root activity, especially in saline conditions. Metabolic activities of roots are unaffected by reducing O_2 concentrations, until they become very low. The knowledge of these low rates – in regular and saline conditions – may help to supply the required oxygen amount, save energy in recycling nutrient solution as in hydroponic water culture method, or in drip irrigation conditions, and add amounts of air or oxygen only, when required by the roots (Leopold and Kriedemann 1975; Woolhouse 1981).

In all culture methods where the roots are attached to a growth medium and require periodic irrigation, a film of nutrient solution remains on the root surface after drainage. The oxygen (air) in this film is available for root respiration and activity. The air between the root growth particles (sand or gravel) is unavailable for root respiration for the most part.

Flushing (in closed systems) or irrigation (in open systems) results in part of the old nutrient solution (with lower oxygen content and higher ion concentration) being replaced by a freshly aerated and more balanced ion solution. Periodic freshening of the nutrient solution around the root surface is of major importance.

Part of the nutrient solution film "hanging" on the growth media particles (such as sand or gravel) flows toward the root surface between irrigation periods and supplies both oxygen and nutrients to the root.

The use of gravelly growth media provides better and quicker exchange of the nutrient solution (including aeration) compared to sandy growth media, but frequent irrigation is required. Sand culture, having less intermedia to space, provides more balanced nutrient solution to the roots, but less solution is exchanged with every irrigation so that oxygen (air) availability to the roots may be limited.

Finding the correct balance between too dry (sufficient oxygen availability to the roots), which may cause partial wilting of the plants, and too wet, which may cause lack of oxygen, is one of the main problems in sand culture as compared to gravel culture.

Roots are the primary absorbing organ for mineral uptake. Certain conclusions have been made on the basis of studies on roots carried out by several independent groups of investigators

1. Salt absorption is usually most rapid in those root tissues where cell growth is by expansion and vacuolation.
2. Root hairs may not be specifically or even more actively associated with salt uptake than any other cells exposed to the same medium (an important fact in soilless culture conditions, where many species have no or only very few root hairs).
3. The site of most active absorption is often 1–2 mm behind the root apex.
4. Older regions of roots with nondividing cells are still capable of active salt absorption.
5. The compulsive pumping of accumulated solutes from roots to other parts of the plant is dependent on a relationship between living cells and does not occur in dead tissues.

Solutes are mobile in both the xylem and phloem. They can be retranslocated in the phloem and in the xylem, respectively, arriving in the shoot via the xylem and in the root via the phloem. Furthermore, the phloem supplies sinks not only downwards from the mature exporting leaves but also upwards in the growing parts of shoots (Luettge 1983).

7.1.1 Oxygen

The root system of a plant needs oxygen to liberate energy from carbohydrates. A full grown tomato plant absorbs 6.7×10^{-3} mg O_2/s on a bright sunny day, and 2.2×10^{-3} mg/s on a cloudy day. During the night it decreases gradually to 2.5 and 1.1×10^{-3} mg/s respectively.

If water is saturated with oxygen, 1 l contains only about 10 mg O_2 at 15 °C and 9 mg O_2 at 20 to 21 °C. A tomato plant in water culture with 5 l of nutrient solution at 20 °C has a reserve of only 45 mg O_2. On a sunny day, this is only sufficient for at most 2 h, if the roots are able to exhaust the reserve. Even if this is locally possible for the roots, the solution must still be mixed and circulated around the roots, as the diffusion rate of oxygen through the water phase is almost negligible. At 18 °C, this is only 1.9×10^{-7} mg O_2/s over an area of 1 cm^2 and a distance of 1 cm for a maximum difference in partial oxygen pressure from 160 mm Hg to zero. This sufficiently demonstrates the necessity of circulating the nutrient solution for water culture and the necessity of intense aeration of this solution (Elstner 1982; Krizek 1982; Gislerod and Kempton 1983; Alwan and Newton 1984).

7.2 Relations of Nutrient Absorption

The accumulation of salts near the root tips is probably due to a high respiratory activity in this region and to the absence of conducting xylem to transport the salts away.

Although the meristematic region absorbs salts, it has little ability to translocate them, so that they are retained there. Older regions, in which root hairs are present and in which the xylem is well differentiated, are far more effective in providing transportable ions.

Most salts are probably absorbed through regions where the root hairs usually appear, and are then moved through the xylem to the upper parts of the plant. In addition, some transport from the root hair region toward the root tip occurs, probably through the phloem or through the apoplast. If salts are entirely absorbed into the apoplast of roots and taken in the xylem without having ever passed through a living cell, the rate at which they are absorbed should be directly related to the water uptake and to respiration. If they are forced to pass through living tissue, a selective membrane could allow relatively more water ions to penetrate, so that salt uptake would not be directly related to water movement through the plant. Many experiments have been performed to determine the extent to which salt and water absorption are related. Different results and thus different conclusions have been obtained (Woolhouse 1981; Luettge 1983; Feldman 1984).

It is usually assumed that in high salt (nutrient) content solutions, the uptake of ions is related to transpiration, and water transport is the limiting factor of ion uptake. In dilute nutrient solutions the absorption process is limited, and not related to transpiration stream.

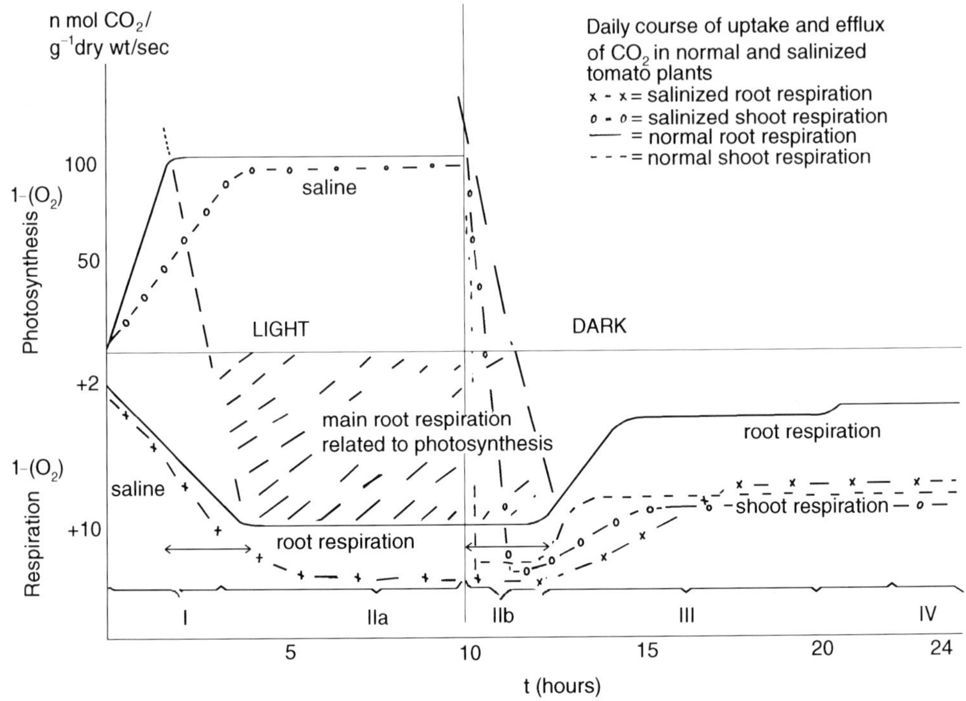

Fig. 7.1. Daily course of uptake and efflux of CO_2. The following periods may be observed in root respiration: Day, *period I*, gradually steep increasing, shortly (min) after starting photosynthesis until (min to h) maximum photosynthesis (related to increased water and nutrient uptake). Day/night, *period IIa*, constantly high, equivalent to maximum photosynthesis. Continuing *IIb* at night (related also to accumulation and storage of photosynthesate). Night, *period III*, gradually steep decreasing, and continuing several hours (related to reduced accumulation and storage of photosynthesate). Night, *period IV*, constant rate

A typical daily course of oxygen requirement for the root in regular and saline conditions is given in Fig. 7.1.

A "typical" daily respiration course of *Xanthium* – or tomato plant – is seen in Fig. 7.1. After light was turned on, root respiration rose. Photosynthesis reached maximum rates after approximately 90 min. The beginning of maximal rate of root respiration is delayed about 130 min. The same delay, of 130 min, is the decline in root respiration after the light was turned off.

7.2.1 Nutrient Uptake and Growth of Nature Cells

Mature cells that have stopped growing are not able to accumulate ions as well as rapidly dividing and growing cells. Young cells actively synthesizing protein create binding sites for such cations as magnesium, iron, calcium, and potassium; while such ions as phosphate, nitrate, and sulfate are rapidly changed into other compounds and also become a part of the increasing protoplasm. This difference is undoubtedly partly responsible for the relation of growth and solute uptake, although other factors may also be involved. It has been suggested, for example, that solute uptake into mature cells not synthesizing protein is decreased because the needed carrier molecules are not being synthesized.

7.2.2 Factors Affecting Nutrient Uptake

Temperature. the rate of salt absorption by plants increases with temperature. The slow growth of plants in cold is partly due to the limited rate of ion uptake under these conditions. Increased absorption with rising temperatures results to some extent from the increased rate of diffusion of salts to the roots and also from the increased respiration of the plants at higher temperatures. The active absorption of elements requires cellular respiration. Ion accumulation is reduced at temperatures above approximately 40 °C. This is likely because such high temperatures interfere with maintenance respiration (oxygen demand) and perhaps also because the cell membranes become more permeable to the passive leakage of salts at high temperatures so that the net rate of accumulation then becomes less. It appears that temperature also affects the relative ratio of cations to anions absorbed. Various studies indicate that at low temperatures the uptake of anions is inhibited more than that of cations, although the reason for this is not clear.

The best root and shoot growth of most vegetables occurs between 15 and 25 °C. The reduction in growth at lower temperatures is attributed partly to water stress caused by reduced absorption of water, decrease in translocation from shoot to root, synthetic activities in the roots, and reduced root differentiation of cells (Schwarz 1972; Marschner 1983; Roeber and Schaller 1985).

7.2.3 Ion Absorption

The absorption of salt by intact plants involves the movement of ions from the root environment to the root surfaces, ion accumulation in root cells, radial movement of ions from root surfaces into the xylem, and finally their translocation from roots to shoots. The ability of cells to accumulate ions to a concentration higher than that in their environment is characteristic of living organisms. One theory of accumulation proposes that anion accumulation is by active transport linked directly to metabolism and that cations move passively along the gradient of anions. A more widely held theory assumes that ions form temporary combinations with special organic molecules called carriers, and that the resulting complex can move through cell membranes which are impermeable to ions alone. The ions then separate from the carriers at the membrane surface and move into the vacuoles, while the carrier molecules become available to move more ions. Metabolic energy is required both for the synthesis of carrier molecules and also to maintain membrane structure. Organic acid metabolism is also involved in maintaining ionic balance.

The kinds and amounts of ions accumulated by root cells vary widely among different species of plants, with the metabolic activity of roots, temperature, and aeration.

How do ions move radially from the root surface into the stele and enter the dead xylem elements? The classical explanation is that ions are accumulated by the cells at the root surface and move inward through the symplast to the stele by diffusion and cytoplasmic streaming. It is not clear how they are transferred from the living cells of the stele into the sap of the dead conducting elements.

Ions which reach the xylem sap of the roots are usually carried to the shoots in the transpiration stream. Some salt is removed from the xylem sap by adjacent living cells. Some inorganic ions, notably nitrate, are converted to organic compounds in the roots. Some mobile ions are moved from old to young leaves, and some ions carried to the shoot in the transpiration stream are moved downward again in the phloem. Thus there is a continual recirculation and redistribution of ions in plants.

The amounts and kinds of ions accumulated vary with the kinds of cells and environmental conditions. Aeration, amount of carbohydrate in the roots, concentration of ions in the cells and in the substrate, pH, light, temperature, and the species of plant are among the factors which affect the rate of ion accumulation.

Nutrient uptake varies diurnally, the peak rate occurring 5–6 h after the peak rate of CO_2 influx, which occurs at about 1200 GMT on a sunny day. Nutrient uptake also decreases quickly after the loss of photosynthetic tissue. These observations imply that nutrient uptake responds to metabolic demand resulting from the production of photosynthate.

The rate of nutrient uptake depends on the external concentration, but this dependence becomes less as plants age. Under steady-state conditions the nutrient flux becomes constant, or nearly so, down to low concentrations when plants are more than about 3–4 weeks old. At very low concentrations the measured nutrient flux can only be accounted for if absorption occurs close to the epidermis.

Under nonsteady conditions the nutrient flux changes quickly after the external concentration is changed. There is an increased flux when the external concentration is increased even though the potential growth rate is achieved at the low concentration (Evants 1975; Luettge 1983, Marschner 1983).

8 Plant Growth Effects and Plant Tests

8.1 Photosynthetic Efficiency

On the average, only one quarter of 1% of the radiant energy reaching the earth is captured by photosynthetic organisms. This largely is a measure of the density of photosynthetic plants on the earth's surface and the efficiency with which they can absorb light and photosynthesize throughout all the seasons. In contrast, a highly efficient C_4 plant, such as corn, may utilize 5% of incident radiant energy during its most rapid period of growth.

What is the potential efficiency of photosynthesis? One answer is provided by measurements of quantum efficiency. Efficiencies of one carbon dioxide molecule fixed per 8 to 10 quanta absorbed have been measured using single-cell algae. These efficiencies are high, in terms of conversion of incident light, and probably can only be achieved under the optimal experimental conditions – very low light intensities in comparison to average sunlight. For this and other reasons, it is impractical to extrapolate these measurements to field crops.

Many factors such as developmental stage of the plant and the presence of biological and physicochemical stresses can reduce photosynthetic efficiency. In addition, each step in the photosynthetic process, from the absorption of light energy to the conversion and storage of energy in the synthesis of sugar molecules, can be affected differently by various limiting factors.

8.2 Harvest Index

The photosynthetic efficiency of a plant depends on the stage of development of the plant and the presence of biological and physicochemical stresses. For farmers, improving the harvest index is more important than improving photosynthetic efficiency.

The most effective way to improve the harvest index (the ratio of harvested part of the plant to total plant) may be to improve the accumulation of photosynthate in the desired plant part.

Traditional breeding methods have been selected for an improved harvest index in many crops and have led to substantial gains in crop yields.

Plants have been selected as crops based on their parts that accumulate the products of photosynthesis. Plants with lush vegetative growth such as lettuce undergo rapid photosynthetic growth. On the other hand, plants that deposit their photosynthate in stems, roots, fruits, or seeds are selected for these attributes and harvested after their storage organs have achieved maximal size. In these plants the attempts are to redirect the photosynthate to the portion of the plant that will be used for feed or food.

Some methods, such as prolonging the plant activity during senescence while most deposits of photosynthate occur, lead to improvement of the harvest index ratio. This can be achieved by proper nutrition balance and environmental control, which will increase the yield. The improvement of root: shoot ratio, and partitioning of energy (carbon) will improve the harvest index. Foliar application of plant growth substances or nutrients can artificially prolong active vegetative growth, and subsequently total crop yield may be increased (Alexander 1986).

Control of the transfer of photosynthate to various parts of the plant and optimization of deposition of carbohydrate, protein, and fat reserves in storage organs determine the harvestable yield of a crop. Relatively little is known about the processes regulating the translocation and metabolism of photosynthetically fixed carbon.

Sometimes environmental factors – such as temperature level at the root-crown area, where in many plants the xylem and phloem cross and form a bottle neck for the upward and downward flow of plant fluids – have an effect (Schwarz 1972). The harvest index is especially influenced by stress factors. Some of these stress factors are interrelated and therefore multiply the stress effect.

The most known stress effects are temperature, light and humidity extremes, salinity, drought, toxic gases and elements, radiation and, of course, unbalanced nutrition and environmental factors previously mentioned. The harvest index and yield are influenced by all these factors.

Some plants, such as soybean and some forms of wheat, destroy themselves in order to produce a viable, energy-rich seed to preserve the plant line for the next season. The production of the grain in wheat plants derives from the flag-leaf in the last stages of its life. If the life of this leaf could be prolonged, higher yield could be expected.

If this delicate control mechanism can be manipulated to prolong the period of active photosynthesis in the plant without destroying its ability to undergo senescence at the proper time, the harvest would be greater.

Maintenance respiration increased with saline and heat stress. Maintenance respiration uses up to 30% of the photosynthesate-produced energy, often leaving insufficient energy for plant products. Increasing the amount of carbon dioxide enhances the photosynthetic capacity under stress conditions and drastically affects the harvest index.

8.3 Root-Shoot Relationships

Shoots are the initial source for most of the organic metabolites used in growth and maintenance, and roots are the initial source for nearly all the inorganic nutrients and for water. There is a close coordination of root and shoot development. When seeds first germinate, much of the metabolite is directed toward root expansion. If half of the root mass is removed from a seedling, root growth will accelerate, relative to shoot growth, until this set fraction is again achieved. Similarly, if half of the shoot mass is removed, shoot growth will increase, relative to root growth, until the ratio again is achieved. Chemical messengers probably are interchanged between roots and shoots to control growth and development processes (Ehret and Ho 1986).

Many factors affect the relative shoot-root proportions. C_4 grasses have lower root – shoot ratios than C_3 grasses. Root growth rate declines during senescence and seed filling, thus decreasing root-shoot ratios. This decline in root growth can be reversed by removal of fruit. Shoot growth rates are increased, relative to root growth, by increasing soil water availability, by water and ion stress, low temperature regime, increasing nitrogen supplies to roots, and by defoliating the plant. Conversely, relative root growth is enhanced after damage to roots; CO_2 concentration is increased around the shoots, light intensity is increased, and soil water content is decreased. Root systems of many kinds of plants respond to waterlogged growth media by increasing their internal porosity through development of large continuous air-filled channels, the aerenchyma. These channels increase gaseous diffusion downward toward the root tips.

Molecular diffusion in response to a concentration gradient is the major source of oxygen for respiration.

Root: shoot ratio increase has been observed with elevated root temperature, under high nitrate concentration in the nutrient solution, and under low total nutrient concentration (Johnson and Ho 1986; Ehret 1985; Mortley et al. 1991).

8.4 Plant Tests: Choice of Tissue

Under conditions of extreme deficiency, small increases in yield that result from the application of the deficient element may sometimes be accompanied by a decrease in the concentration of the element in the dry matter.

Many different plant organs and tissues have been used for diagnosis. The tissues include roots, stems, bark, leaves of different age and position, laminae, petioles, midribs, seed, fruit, and grain. In general, changes in nutrient concentration, caused by varying the nutrient supply, are greater in leaves than in other organs. Although the entire leaf is usually sampled, the concentration in the petiole dry matter may occasionally be very different from that of the laminae, so that the inclusion of variable lengths of petiole will give variable results. Occasionally, the petiole will give a better indication of the nutrient status than the lamina. This, however, may be true of one element but not of another.

Change in the supply of nutrients can markedly affect plant morphology, thereby altering the proportion of dry matter distributed to different organs. It is important that this does not distort the correlation between nutrient content and yield or growth. There are no general recommendations and each crop, element, and situation requires its own extensive study in order to develop a reliable diagnostic system based on plant analysis.

Within a given environment, the nutrient uptake by a plant depends on the supply to the roots and on the demands for the nutrient set up within the plant by the growth and normal functions of plant organs, including the roots.

Each vegetative organ of a plant passes through phases of intake, a relatively constant content, and then export of many of its mineral nutrients. Each organ therefore has a certain capacity to accumulate nutrients, and may subsequently during senescence constitute a potential source of nutrients for younger plant parts. The demand of each organ for nutrients is usually met in part, if not entirely, by uptake from the medium, but the rate of uptake depends to a certain extent on the availability of the nutrient element within the plant.

In annuals, with their shorter period of growth and with a more rapid development, changes in nutrient content tend to be more rapid and extensive. It is generally difficult to use plant analysis to remedy current nutritional disorders, unless the diagnosis is made early enough in the life of the plant (Ward 1973).

In the early stages of growth of annual plants the tissue concentration of nutrient elements tends to be high. As the plant develops, nutrient uptake continues, but for some elements, particularly mobile ones like N and P, the concentration in the plant tops or in the whole plant as a percentage of dry matter usually declines. The process is usually referred to as dilution and is partly due to an increase in the proportion of structural and storage materials relative to protoplasm. Other factors can be a decline in the nutrient supply to the roots, in an absolute sense or relative to the rapidly increasing demands for growth. In the case of mobile elements, there is also the redistribution between organs initiated by internal competition for nutrients or by senescence, even at adequate levels of external supply.

Not all elements are mobile in the plant: for example, Ca is not. Plants transferred from solutions containing luxury Ca levels to Ca-deficient solutions developed Ca-deficiency symptoms, even though the plant tops had Ca concentrations three to ten times higher than the tops of healthy plants grown in a low but constant Ca supply.

These considerations apply to any element with a restricted mobility in the plant. The difficulties in diagnosing deficiencies of these elements can be lessened by avoidance of old leaves and a careful selection of young tissues, e.g., leaves or even roots.

Nutrient interactions, involving a change in the concentration of one element in plant tissue caused by another, are quite common. They are often difficult to unravel, but need to be taken into account in diagnosis based on plant analysis. A shift in the content of one element invariably is accompanied by secondary changes in tissue content of other elements, even though there is no change in the availability of the interacting nutrient(s) (Emmert 1961). Alternative expressions for the same phenomena are synergism and antagonism (Smith 1962). Antagonism can occur during ion uptake, during translocation and accumulation in the tissues, or in metabolism (e.g., in enzyme systems, Fe versus Mn; Prevot and Ollagnier 1961). It can involve competition between two or more elements, but also precipitation of nutrients or other phenomena.

Antagonisms during uptake may occur between cations but also between anions. Well-known examples of cation antagonism are those between K and Ca, K and Mg, and Fe and Mn. N supplied as the ammonium ion is antagonistic to other cations, but when supplied as nitrate it may compete with other anions, e.g., phosphate. Increasing the level of K to fruit trees has been shown to induce Mg deficiencies, accompanied by increases in the K concentration of the leaves. Conversely, changes in Mg concentration of leaf tissue are usually accompanied by opposite changes in leaf K. Antagonistic effects have also been extensively documented for micronutrients. High levels of Cu can induce Fe deficiency; high levels of Cu, Zn or Mn can induce marked antagonistic effects between pairs of these ions on concentrations in leaves (Marschner 1983).

Antagonism during translocation is often caused by precipitation in the root tissues or elsewhere in the plant. Excess phosphate in the root environment causes precipitation of Zn and also of Fe along the veins, and this may lead to iron chlorosis.

Other factors affecting nutrient compositions include the fruit load and the position of the leaves relative to the fruit. Temperature, light, and moisture not at the optimum level may also influence the nutrient content, as can poor root aeration. Plant diseases may be misleading interpretation of tissue composition.

8.5 Plant Analysis

To send samples for plant or leaf analysis, a specific plant part is taken from a number of plants and submitted to a laboratory for analysis and evaluation.

The objective for using plant analysis, as a part of the growing routine, is to determine if the nutrient solution management technique is providing all the essential elements in their proper ratio and concentration, and so warn of possible elemental stress before an insufficiency or toxicity actually occurs. This procedure of routine sampling and analysis is then referred to as tracking. It is done to provide the information needed to establish the proper nutrient solution management procedures required to ensure that all of the essential element levels are within the sufficiency range for the crop being grown. It is well worth the time and expense to develop a track of elemental sufficiency, that can be used to firmly establish the proper nutrient solution management system for future use (Ward 1973).

A grower, faced with a suspected essential element deficiency or imbalance, should verify the suspected insufficiency by means of plant analysis. Many symptoms of elemental stress are quite similar. The stress may be due to the relationship between elements, and so require more than just a simple change in the nutrient solution formula to correct. Without an analysis, an incorrect change may be made which could further aggravate the problem.

Since plant analysis requires the use of a competent laboratory, contact with the laboratory should be made before samples are collected. The laboratory will provide sampling and submission procedures that are important to follow.

If no specific sampling procedures are provided for a particular plant, or the time period is not specifically provided, the rule is to collect recently mature leaves below the growing tip. This sampling procedure should be followed if the plants are being monitored in order to develop a track of their elemental content.

Collect a representative sample by taking one leaf from each of eight to ten plants chosen at random in the test area (usually one sample per greenhouse). For tomatoes, the standard leaf for tissue analysis is the fifth leaf from the growing tip of the main stem. Tomato plants cannot be sampled after the plants have been topped.

For cucumbers, the sample leaf is usually the third visible leaf from the top, and should be about 10 cm in diameter. After topping, select a similar leaf from the highest lateral on the plant.

Cut off the leaf close to the main stem. With tomato, send in the whole sample for analysis. With cucumber, cut off the petiole close to the leaf and discard it; send only the leaf portion for analysis.

Wrap the leaves in a paper towel and place the towel in a plastic bag. Press as much air as possible out of the bag and tie with a “twistem” or string.

When visual symptoms of plant stress are evident, it is advisable to take similar plant tissues from both affected and normal plants. In this way a comparison of difference is made which can be far more helpful than just an analysis of the stressed plants alone.

Plants which are diseased, insect- or mechanically damaged should not be selected for sampling. Dead plant tissue should not be included in the selected sample. Tissue that is dusty or coated with chemicals should be avoided.

Once the tissues have been collected, it is best to air dry them before sending them to the laboratory for analysis. This will keep them from rotting while in transit.

Quick testing kits for plant analysis are available from some chemical testing companies. The use of these tests is recommended only when laboratories cannot be consulted frequently.

8.6 Disease Control

8.6.1 Sanitation of Diseased Plants

Diseased plants should be treated as in soil culture. Affected plants or plant parts should be removed immediately to avoid infesting other plants. Diseased plants should never be ignored in the hope that they will provide normal yields.

Most disease control sprays will not affect nutrient solutions, but when using copper- or sulfur-containing sprays, care should be taken not to leave large amounts of these elements on the media.

To determine whether a disease-like symptom is a result of a nutrient deficiency, spray the specific nutrient on some plants to ascertain if the hypothesis is correct. Addition of a wetting agent, as in spraying for disease control, is recommended.

Spraying is most effective in early growth stages or on young leaves. A 1–3% macronutrient and a 0.1–1% micronutrient solution are generally recommended. A nutrient spraying, especially a micronutrient spraying, should be effective in 1–3 days. When the result is positive, the missing element should be added to the nutrient solution or the nutrient solution should be replaced. A nutrient-element spray can prevent or control nutrient deficiency only temporarily.

8.6.2 Root Diseases

Closed hydroponic systems and NFT systems permit a rapid spread of root diseases. With tomato, the most common crop, some natural root death occurs when the extension growth of the roots ceases as the first fruits begin to swell. At that time of natural stress, the plant is vulnerable to root pathogens, which may be introduced by contaminated seed, infected propagating mixtures, or even adjacent greenhouse soil.

Common root pathogens causing tomato wilts are *Fusarium* and *Verticillium* spp.; these diseases are most easily avoided by using resistant cultivars.

The use of ultraviolet radiation to reduce bacterial plant pathogens in nutrient solutions has appeared promising in some NFT systems.

No chemicals have been found which could be added to the nutrient solution to destroy or avoid development of root pathogens.

8.6.3 Sterilization of the Growth Medium

Sterilization of the growth medium after each crop is recommended. A good sterilant will destroy bacteria, fungi, and viruses, providing a disease-free medium for the new crop. This subject was partially covered under the discussions on the various soilless-culture methods.

Many chemicals are available for sterilizing the growth medium. The best and quickest acting is methyl bromide, a gaseous fumigant, supplied as a liquid under pressure in metal cylinders. About 20 g methyl bromide per m^2 is recommended. It should not be used when the next crop is to be carnation.

A number of chlorine compounds are useful sterilants for small areas. Calcium or sodium hypochlorite (3000 to 10000 ppm) solution or sodium chlorophenate (Dowicide, 300 ppm) may be used. The solution should be left for 24 h, then washed out three or four times. Beds may be used after 38 h. Local chemical consultants should be involved in determining the best available material.

When using subirrigation methods, the medium should first be cleared by filling the beds with water and removing all roots before sterilization. Most small rootlets will be washed out by letting the water overflow. After 4–5 years of use, it is advisable to remove all the gravel and clean the beds.

In some very hot climatic conditions, a good sterilization can be obtained by laying black plastic over the growth media for a week. The growth media will heat up enough to destroy all living material underneath.

The fear of root disease, even if restricted, is the main reason that closed soilless culture systems have decreased in popularity. Due to recent environmental restrictions in some countries, such as The Netherlands and Germany, to prevent contamination of the groundwater by percolation of the used nutrient solution in the soil, and to the requirement to reuse the nutrient solution, closed soilless culture systems are making a comeback. New sterilization methods such as ozone are under investigation.

The beginner in hydroponic farming often tends to blame his failures on the nutrient solution composition instead of on his ignorance of plant protection, etc. It is thus advisable first to look for mechanical or climatic damage, parasitic attacks and diseases, and to investigate irrigation practices before looking for nutrient disorders. Appendix Table 17 should assist in interpreting the relation of certain common symptoms to their possible causes. The specific symptoms are discussed in Chapter 3 in connection with the various ions; see also Chapter 7.4.

9 Controlling Systems

Beside the nutritional control, the productivity of greenhouse crops depends on the proper control of environmental factors such as temperature, relative humidity (RH), carbon dioxide levels, light, and irrigation.

Some plant growth factors are both environmental and nutritional, such as carbon dioxide and water. Control of aboveground environment is different from that of belowground environment, and the optimal rate for plant growth of the various growth factors differs. Aboveground growth factors can be affected and are related to belowground growth factors and vice versa. These interrelations affect the ability to control growth precisely.

As already mentioned, most growth factors have relatively large ranges for optimal plant growth, and the plant growth controllers are adjusted to these ranges. There are usually three components in the regulation of any growth factor:

- Controller
- Sensor(s),
- Process equipment (PE).

The sensor measures the actual level of a given parameter (temperature, CO_2, RH) and gives this signal to the controller, which in turn determines whether the process equipment (ventilator, mixing valve, etc.) is to be activated and for what time span. An important requirement is that the actual value is achieved with minimal operation of the process equipment (Alberry et al. 1985).

9.1 Controllers

The controller is the mid-point and critical decision-maker in the regulation of the growth factors. An explanation of different controllers is given below the air temperature (T_a) will be taken as an example:

On/off controllers: these units have two signals, namely, for turning a heater "on" when the temperature is too low and "off" when the temperature is too high. Most thermostats have a temperature differential (TD) which can be set manually. A small TD means that the heater runs frequently but in short intervals, and a large TD means that the heater runs infrequently but for longer periods. The fluctuation of the temperature also depends both on the size of the heating system as well as outside conditions, and the TD has to be set for each specific condition(s).

Modulating controllers: these are used to control mixing valves in a "pulsating" fashion to prolong the time of correction and to prevent overrides.

Proportional controllers: these units provide signals depending on the deviation of the setpoint. The larger the deviation from the setpoint, the larger the correction will be (opening of vents/mixing valve). These controllers are often used with mixing valves or pressure regulators.

Microprocessor controller: to accommodate a system which can act as a controller and interacts with all the different options such as light-dependent, time-dependent, minimum and maximum control of the process equipment, interaction of the different parameters, etc., a micropro-

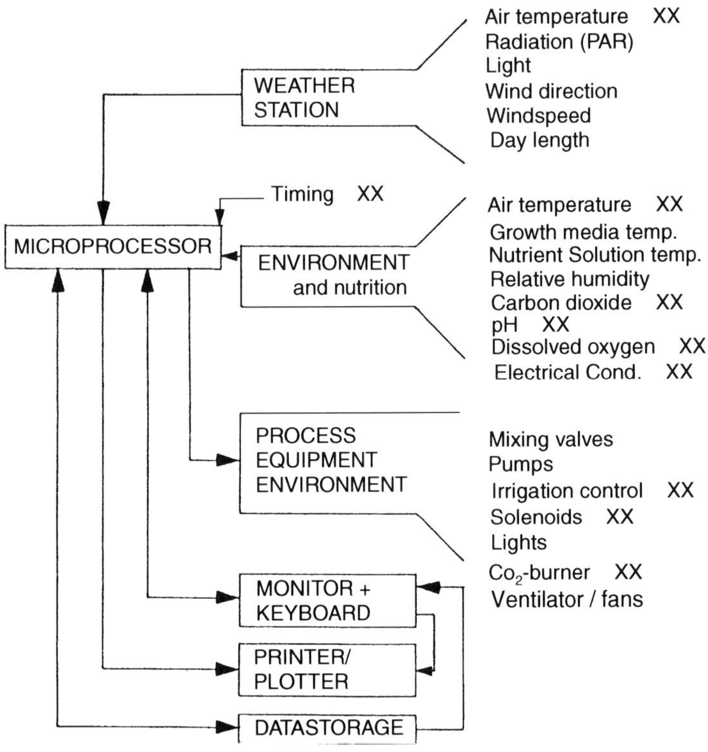

Fig. 9.1. A schematic view of a microprocessor controller. Most common soilless culture microprocessors for open culture systems always include items marked *xx*, and often include items marked *x*

cessor must be considered. This unit can perform many control functions simultaneously for many environments while the outside conditions are measured only once per sensing cycle. The sensing cycle varies from computer to computer type and also with the control function it has to perform. For climate control a sensing cycle of 20–30 s may suffice while the control of the electrical conductivity and pH in the irrigation cycle may require a sensing of every 10 s. A microprocessor also provides an opportunity to store or print data so that actual values can be compared with the set values.

The air temperature can be variable relative to sunrise and sunset as well as light-dependent and/or dependent on the total daily or weekly radiation sum. The water temperature coming from the boilers as well as entering the heating lines can also be light, ventilation, RH and/or minimum- and maximum-dependent. If two or more heating groups are installed, each heating system can be independently controlled, so that the systems can run either in series, in parallel, or separate from other heating groups with their own set point(s).

A schematic view of a microprocessor is given in Fig. 9.1.

9.1.1 Sensors

For each sensing element, accuracy and a short response time are necessary, as most parameters change continuously and sometimes very rapidly (injecting fertilizer in irrigation water). For each parameter there are several devices:

Temperature: the mechanical systems include a bimetal (copper/steel), a liquid or a liquid vapor thermostat. All of the above three types of thermostats involve the principle that a material expands with increasing temperature.

Other types of sensors used in electronic controllers are the ones which change resistance depending on temperature. The most common ones are thermistors. Platinum sensors are often used to measure water temperature within a pipe.

Thermocouples are sensors which are made of two conductor wires which are fused at the end. Two wires of different metal are used (e.g., copper/constantan). A voltage difference between the two leads gives an indication for the temperature. In most cases, correction must be made for the resistance in the wires connecting the sensors with the controlling/sensing device.

Relative humidity: the length dependency of a bundle of hair on the water vapor pressure in the air is often used to measure the relative humidity. This can be through a mechanical means or by changing an electrical resistance. The greenhouse environment with the dust, grease, pesticides, etc. causes a rapid aging of this type of sensor. Continuous cleaning and recalibration is therefore a requirement. Other methods include the measurements of a dry and wet bulb temperature (psychrometer). This method involves a temperature sensor covered with a cotton wick which is kept wet constantly from a distilled water reservoir. Forced air movement (2 m/s) along the wick caused evaporation from the wick and a reduction in temperature. The degree of temperature reduction relative to that of the dry bulb sensor is used to determine the RH. Psychrometers are used by most manufacturers for microprocessors of greenhouse climate control.

Carbon dioxide: the principle of measurement is based on the fact that carbon dioxide molecules absorb infrared radiation of a given wavelength range. Air samples from the greenhouse are drawn through the measuring chamber and compared to the minimum setting. Multiplexers in combination with one sensor are often used to measure the CO_2 level of several compartments to offset the cost of the infrared sensing device.

Light: there are three types of light sensors commonly used in the horticultural industry. The differences are based on the wavelength range measured.

Light sensors: these sensors measure the light photometrically, having the highest sensitivity in the yellow spectrum and tapering down in sensitivity towards the blue and red wavelength regions. The measurements are given either in lx or footcandles (1 footcandle = 10.76 lx).

Radiation sensors: the principle of variable thermal response to radiation of differing wavelength is the basis of radiation measurement. Meterological stations around the world use Kipp and Zonen sensors to measure the global radiation. These sensors determine incoming solar radiation in the wavelength of 0.2–15 microns. The units are usually in W/m^2 or $J/cm^2/day$. Only part of this radiation (about 50%) is within the visible range (photosynthetically active radiation, PAR).

Quantum sensors measure the energy received between 0.4 and 0.7 μ, and imitate the sensitivity of the plants for visible radiation in terms of photosynthesis. As a result, a good correlation has been found between plant growth and quanta. The units are in microEinstein per m^2 per second ($\mu E/m^2/s$).

The sensing elements for temperature and relative humidity should be protected from the influence of outside radiation and unregulated air currents. Therefore, the sensing elements are to be located in an aspirated box in an appropriate location in the greenhouse where plant growth takes place. A small fan draws air continuously through the box to provide a more representative sample. The carbon dioxide sampling tube(s) should be located within the crop canopy, while light sensors are usually located outside the greenhouse in a nonshaded area.

Process equipment (PE): this is the electrical equipment required to make the changes necessary as indicated by the controller. This equipment includes, ventilator(s), mixing valve(s), fans, carbon dioxide generators, dehumidifiers, solenoids, etc. The capacity of the PE and the rate at which it operates determines the response time of a given parameter. Slow moving ventilators or mixing valves allow time for the sensors to interreact with the changes occurring within the environment.

An effective control system should be equipped with overrides and alarms. The overrides can be a manual or an alternative automatic system. If a mechanical malfunction occurs, the systems should have the capability of manual operation. Alarm systems are often combined.

9.2 Temperature

For most plant species and even for individual tissues, the Q_{10} for respiration between 5 and 25 °C is usually near 2.0 to 2.5. With further increase in temperature up to 30 and 35 °C, the respiration rate increases, but less rapidly, so that the Q_{10} begins to decrease. A possible explanation for the decrease is that the rate of O_2 penetration into the tissue begins to limit respiration at these higher temperatures where chemical reactions could otherwise still proceed rapidly. Diffusion of O_2 and CO_2 are also stimulated by increased temperature, but the Q_{10} for these physical processes is less than for the chemical reactions of respiration.

With further rise in temperature, respiration is actually decreased, especially if the plants are maintained under such conditions for long periods. Apparently the required enzymes begin to be denatured at a rapid rate, thus preventing a continued stimulation of metabolism. The actual range of temperature at which respiration occurs in various plants is surprisingly large.

Molecules absorbing visible light are colored or black (pigments). The electrons that become excited are usually particularly mobile electrons associated with unsaturated double bonds. Clorophyll, for example, has a high degree of unsaturation and efficiently absorbs light, especially blue and red. The blue light possesses more energy per photon than does the red, and so it raises the electron to a higher energy level.

The excitation energy induced in a molecule or an atom by a photon may be lost in any of three ways. First, it may be immediately lost as heat (translational, vibrational, or rotational energy). Second, it may be partially lost as heat, with the remainder emitted as visible light of a wavelength longer than absorbed. This is called fluorescence. When fluorescence occurs, is usually does so very rapidly, generally within 10^{-9} to 10^{-7} s after absorption of the original radiation. Third, the energy may be used to cause a chemical reaction. Photosynthesis occurs as a result of this third process. Fluorescence of chlorophyll can also readily be observed as the red light emitted when a concentrated solution of the pigment is illuminated with white or ultra violet light. Fluorescence of the chlorophyll in a leaf is not as readily seen, because most of the excitation energy is instead used to cause photosynthesis.

Optimal temperatures for maximum growth processes vary according to the different plant species, shoot and root environment, and time of day, and are dependent on light intensity and humidity.

The temperature range over which plants can photosynthesize is large. The optimum temperatures for photosynthesis probably fall within 20 and 35 °C for most species of temperate regions. It is often observed that maximum photosynthesis occurs at temperatures a few degrees lower than those for most rapid respiration.

The influence of temperature on photosynthesis depends mainly on the light intensity and the CO_2 availability. If these two factors are adequate, photosynthesis will occur faster at moderately higher temperatures.

High temperatures increase transpiration, as well as the rate of respiration. These processes require energy, and the carbon balance in the plant is affected, which may influence photosynthesate translocation and accumulation. This will influence the rate of development of leaves, flowers, roots, and fruits.

It has been observed that temperature has an influence on the size and the shape of leaves. For example, cucumbers and lettuce have short, wide leaves when the temperature at initiation is low, while at high temperatures the leaves are long and narrow. The difference between night and day temperatures also has an effect on stem elongation. A higher day than night temperature results in a taller plant than one grown with lower day than night temperature. Plants grown at 22/16 °C (night/day) are shorter and stockier than those grown at 16/22 °C/. This can be important both in the plant-raising and production stages of tomatoes or cucumbers. Increase in temperature decreases the relative humidity (RH). Plants grown at high RH have larger leaves and longer stems.

9.2.1 Temperature Sensing and Control

Electronic sensors (thermistors) are used in conjunction with analogue or digital controllers. The sensor should measure the actively growing part of the plant, and therefore be placed near the plant canopy. It is sometimes advisable to have multiple sensing units within each greenhouse or to have the thermometer placed in an aspirated box. The aspirator will draw greenhouse air across the thermometer and give it a more representative sample of temperature.

By using computers to maintain the greenhouse, the temperature can be automatically controlled, depending on the time of day, intensity of sunlight, relative humidity, or many other factors. In addition to a static temperature set point, the computer will also allow for the ramping of temperatures from night to day. Ramping (controlled temperature increase or decrease) will provide more accurate optimum temperatures for plant growth. Although computers require an expensive initial investment, they can be economically viable for large greenhouse operations or in cases where many different temperatures are being maintained.

In glasshouses, using closed circulation soilless culture systems, when reservoirs are used for storing nutrient solution, this solution could be warmed and the root environment thus temperature-controlled. The sensing could be combined with the air environment sensor facilities. A similar solution for warm roots could be achieved by warming the nutrient solution by drip irrigation methods.

If a short 1–2-h frost is expected, the beds may be irrigated with warm solution to increase the air temperature around the root crown and the lower plant parts.

In outdoor installations, using dark growth media, the media surface temperature may be up to 7 °C above that of the air or that at 2–3 cm depth. This can cause burn to young unshaded plants, especially at the root crown. These temperature differences are much less marked with light color media. Dark media absorb more light energy and will, therefore, promote seed and seedling development in water. The growth media may be warmed by electricity, with the wires under or on the growth media. This method is very expensive and seldom economic even in greenhouses or when plastic covers are used. A cheaper method of heating the media is by warming the irrigating nutrient solution.

9.2.2 Heat Suppliers

Many books have dealt with this subject, and following are some remarks on nonconventional energy-heat options.

Waste heat: Probably the most feasible sources of low-cost thermal energy are industrial plants that produce large amounts of waste heat, principally steam-electric generating stations. Such heat is released from condensor cooling water to the atmosphere via cooling towers, or the warmed water is returned to its source, usually a river, an impoundment, or the sea. An estimated 2 kWh of waste energy are produced for every kWh of electrical power produced.

Because such generating plants are operated to provide electricity, the amounts and temperatures of cooling water may vary somewhat. However, large base-load plants cycle out so much heated water that only a fraction of the rejected heat could be utilized at best; and though condensers and mechanical systems must be cleaned and chemically treated periodically, the changes in water quality caused by such processes can be avoided by the use of heat exchanges. The cost of heat-exchanger systems is an inhibiting factor.

Waste heat from nuclear power reactors could be an important source for reducing energy-heat expenses.

Solar collectors, associated with heat transportation and storage systems, seem to be the most convenient. However, the required collector area is approximately 3/4 that of the greenhouse itself.

Geothermal energy is used in some locations. The heat exchange and distribution systems are relatively expensive. In some installations the hydroponic nutrient storage tank is the heat storage.

9.3 Light

Light is the primary productive factor in plant growth, through photosynthesis, and is often the main limiting factor. The amount of light, the spectrum of the light and the duration of the light (day/night) are the main parameters. In open areas, the grower has to adapt the crop to the seasonal changes in the light and diurnal periods.

Shading under subtropical conditions may provide better conditions for young seedling production. Adequate shading reduces light intensity by 25–50%, leaving enough light for optimal plant growth. Fruit crops, such as tomatoes, generally need more light than vegetative crops, such as lettuce. Most cut flowers will be of poorer quality if unshaded in summer time. Leaves may appear slightly darker green in dim than in bright light because the chloroplasts are then oriented closer to the leaf surfaces.

9.3.1 Effects of Light on Chloroplast Arrangement

Light affects the arrangement of chloroplasts in the cells of many plants. Chloroplasts are found in the cytoplasm and not in the vacuole. When light intensity is strong, as in full sunlight, the chloroplasts are aligned along the radial walls, becoming shaded against damage by strong illumination. In weak light, or sometimes in darkness, they are separated into two groups distributed along the cell walls nearest and farthest from the light source. More of the low-intensity light is then absorbed. In darkness, they are also often scattered between the two positions mentioned.

9.3.2 Light Intensity and Plant Leaf

There is an apparent interaction between leaf size and light intensity. This is particularly true for young expanding leaves during periods of low light intensity. Experiments have shown that a reduced light intensity during a certain period is compensated to some degree by a larger leaf surface area, without having an effect on the rate of photosynthesis of these leaves per unit surface area. Larger leaves also mean that light will be better intercepted by the plant to improve the overall conversion efficiency of the plant. The leaves are usually thin and weak. Leaves formed under high light intensities usually have a thicker cuticle, with the chloroplasts being arranged parallel to the sun radiation (inefficient) compared to leaves formed during light intensities (chloroplasts perpendicular to sun radiation). Although plants have the ability to adjust to lower light levels, it depends on the age of the plant (leaves) and the length of time of acclimatization. The above reasoning explains for a part why spring-grown plants are bigger than summer-grown plants even though the same amount of light may be received by both.

Effect of the Crop Canopy. Light is being reflected, absorbed, and transmitted by each leaf and only a fraction of the light energy is converted into carbohydrates. The use of several leaves on top of each other will make it more efficient for a crop as a whole (up to 10%). The leaf area index (surface area) of young leaves is usually more efficient than of older ones in terms of net photosynthesis (gross photosynthesis minus respiration). Growers use that to their advantage by keeping the younger leaves on top (tomatoes and cucumbers) in order to benefit most from the higher light intensity in the top of the canopy.

Although there is a direct relationship between light intensity or a light duration for a single leaf, this is not the case for a whole plant. The effect of the changing environmental factors (temperature, RH, and CO_2) during the season have also a definite effect on the total outcome as well (Tremblay et al. 1984).

9.3.3 Diurnal Light Fluctuation

Figures 9.2, 9.3, and 9.4 explain the dependence of plant activity on the diurnal light radiation.

9.3.4 Light Duration and Intensity

To calculate the average photosynthetically active radiation (PAR) within the greenhouse, it is assumed that about 50% of the total radiation is received in the visible spectrum (0.4 to 0.7 μ) and that about 60% is received at plant level in the greenhouse. It should be remembered that these numbers represent the average for the light intensity by dividing the available light by the number of hours of day length. Actual light intensities increase from sunrise with its maximum at zenith (about noon) after which it decreases to sunset.

During the winter, light is often the limiting factor for greenhouse production. However, the relationship between production and light for greenhouse production is not simple.

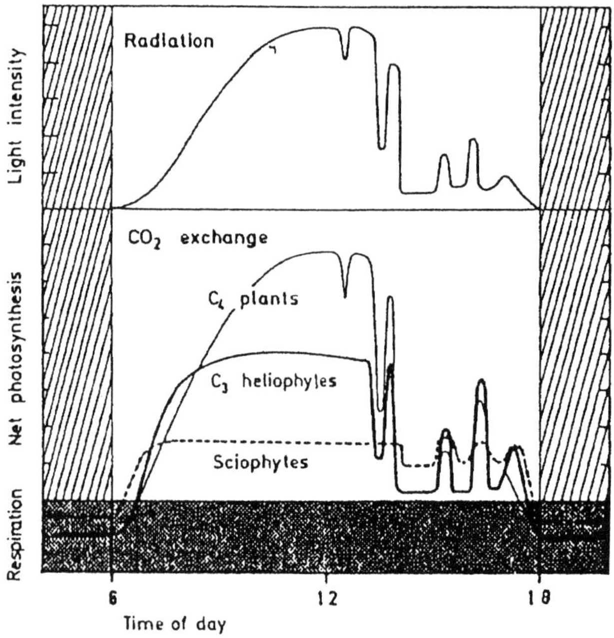

Fig. 9.2. Diurnal light fluctation. Schematic diagram of the daily fluctuation in CO_2 exchange as a function of the available radiation. C_4 plants can utilize even the most intense illumination for photosyntheses, and their CO_2 uptake follows closely the changes in radiation intensity. In C_3 plants photosynthesis becomes light-saturated sooner, so that strong irradiation is not completely utilized. Sciophytes, adapted to utilization of dim light, take up more CO_2 in the early morning and late evening, as well as during periods when the sun is obscured, than do the heliphytes; but the former do not utilize bright light as efficiently

The rate of assimilation is directly related to the duration of the light period, i.e., an 8-h light period will produce twice the amount of assimilation products compared to a 4-h period, provided the light intensity is constant. The law of diminishing return is applicable when the rate of assimilation is related to the amount of photosynthetic active radiation, i.e., the advantage of the first 10 W/m^2 is higher than the additional advantage obtained from increasing the light intensity from 20 to 30 W/m^2. This means that the light conversion efficiency is higher for low light intensities during the early morning hours than at noon. This is illustrated in Fig. 9.4.

9.3.5 Light in Glasshouses

The light in glasshouses is often the limiting plant growth factor. The amount of light which is received inside the greenhouse is strongly

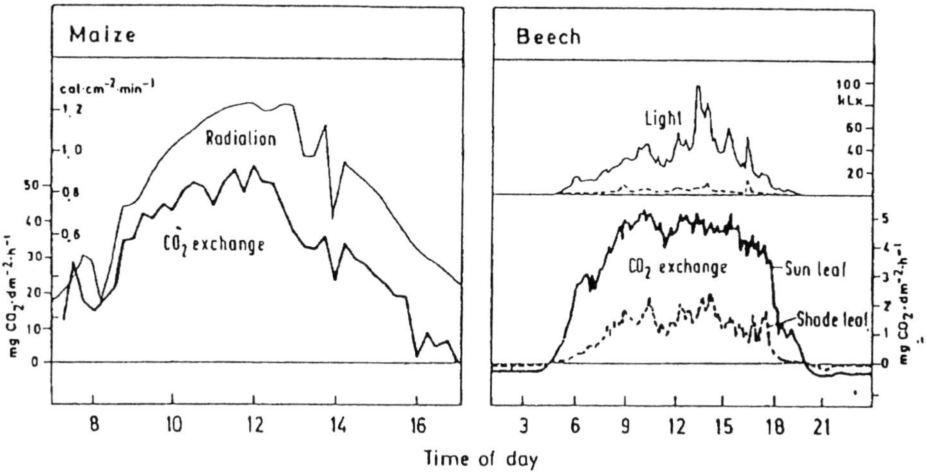

Fig. 9.3. Variations in net photosynthesis measured throughout the day as a function of the available radiation. In maize (a C_4 plant), photosynthesis follows the daily fluctuation in radiation. In sun leaves of the beech, F_n follows the changing illumination only up to about 50 klx, and in shade leaves, F_n undergoes short-term fluctuations associated with the brief fluctuations in brightness

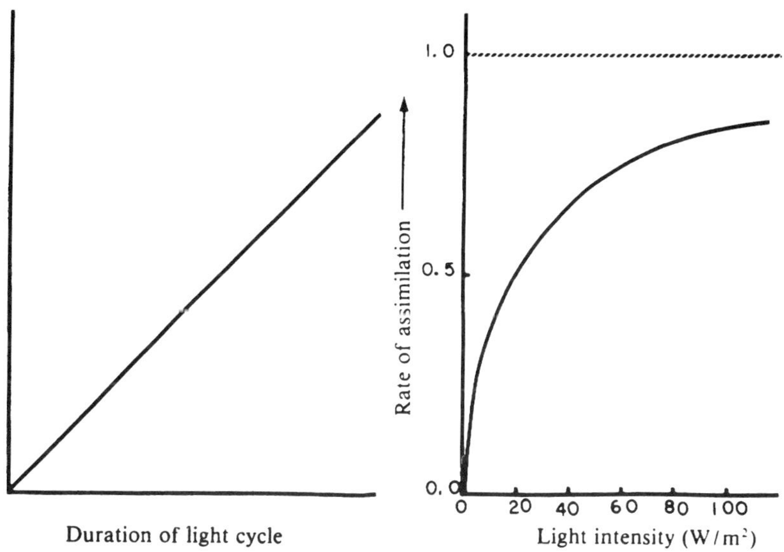

Fig. 9.4. Dependence of carbon dioxide assimilation on duration of light cycle and light intensity. The relationship between light period or light intensity (PAR) and rate of assimilation. (According to Drs. O. Elgersma and Philips, The Netherlands)

dependent upon the glazing material. Glass has been the traditional glazing for vegetable greenhouses. However, there has been a notable shift to the use of double-layers polyethylene for new greenhouses. The reasons for this change are energy conservation and low capital construction costs. Current designs of polyethylene greenhouses provide for optimum but somewhat reduced light transmission, when compared with new glass greenhouses.

Many conflicting reports have been published on light transmission of different materials. Different measuring techniques usually account for these discrepancies. Light transmission is defined as the ratio of light passed through a material and the original amount of light striking the outer surface of any object when that light is directed perpendicular to the material. For plant growth, the light between 400 and 700 nM wavelengths is the most useful for photosynthesis. Light intensity is often expressed in radiation terms as watts per square meter (W/m^2) or photometrically in lux or foot candles.

Two major factors which determine the light transmission of glazing are the refractive index and the amount of light absorbed by the material itself.

The light transmission of a regular single-layered glass material is about 90%, and a double layered 80%. Mainly the amount of sheets (layers) and the thickness of the material determine the amount of light reduction. The major loss of light intensity is through reflection, not absorption.

Incandescent (tungsten-filament) lamps are generally not used for supplemental lighting because of excessive heat. Light quality is acceptable for some plants but insufficient at the low intensities needed to avoid excess heat. For other plants, the high proportion of red and far-red light emitted caused tall, soft growth and other changes in plant form. These lamps are also very inefficient, converting only 7% of the electrical energy consumed into light energy. Much of the energy is converted into heat, which at times is of value and at other times a detriment.

Fluorescent lamps are used more commonly in growing rooms and over small germination areas rather than in full-sized greenhouses.

Among the more efficient of the fluorescent lamps are the cool white and warm white tube types. These convert 20% of the electrical energy consumed by the lamp and the ballast to visible light energy and have similar spectral light emissions. Cool white is perhaps the most commonly used fluorescent lamp for plant growth. The light emitted tends to predominate in the blue region, There are a number of other fluorescent lamps with special phosphors for emitting a spectrum of wavelengths more in line with the requirements of photosynthesis.

A limiting factor to the use of fluorescent lamps for supplemental lighting is the lower power (wattage) of tubes available. This increases the number of tubes and the total cost of fixtures and wiring needed. The larger number of fixtures increases the area of shadows cast on the crop.

High pressure sodium (HPS) lamps are used in many greenhouses. The light emission of spectrum predominates more in the higher wave lengths with a peak of 589 nm (yellow). The light-emission spectrum extends beyond the visible range (400 to 700 nm) into the 700- to 850-nm range. Radiation in this latter range is required for stem elongation, increased fresh weight; and early flowering of most plants. HPS lamps are very efficient, converting 25% of the electrical input into the lamp and ballast into visible radiation (400–700 nm). Suitable models of HPS lamps for greenhouse use are available in 100- and 1000-W sizes. The life expectancy of HPS lamps can be as much as 24000 h

Low-pressure sodium (LPS) type HID lamps are the most efficient of all lamps used for supplemental greenhouse lighting. Twenty-seven percent of the electrical input into lamp and ballast is converted to visible radiation. There is almost as much usable light energy from a 180-W LPS lamp as from a 400-W high-pressure mercury lamp. These lamps have a life expectancy of 18000 h.

LPS lamps emit most of the light in the band around 589 nm, and little in the 700 to 850 nm range. About 10% of the total light is supplied from incandescent light. This is especially important in lettuce plants and some flowers.

In the northern latitudes, the main growth limiting factor in winter time is light. Supplemental lighting is costly. The installation cost of a lighting system may come to US$ 50/m^2, and the yearly operating cost to US$ 15/m^2.

9.4 Automated Fertilizer Application

The most expedient method for applying fertilizer is the automatic watering system present in most greenhouses. The fertilizer must be dissolved into a concentrated solution in order to conserve space in the mixing and holding tanks. This necessitates the use of an injector, otherwise known as a proportioner. This device mixes precise volumes of concentrated fertilizer solution and water together. By plumbing the proportioner into the main water line, serving a greenhouse range, all lines will carry a single-strength fertilizer solution. The proportioner is located on a bypass line so that either water or fertilizer solution can be obtained from the lines.

It is advisable that in a potable water system a backflow preventer be installed on any water supply fixture that has an outlet which may be submerged. Such fixtures include fertilizer proportioners and hoses used to fill spray tanks or equipment washtubs. The backflow preventer stops backsiphoning of contaminated water into the water system in the event that a negative pressure (suction) develops.

Fertilizer proportioners: an automatically controlled open system utilizes fertilizer proportioners, manufactured devices that inject specific amounts of nutrient solution into the irrigation water. For such usage, the solution must be highly concentrated and is prepared in two separate mixtures: one containing calcium nitrate and iron, the other containing the balance of dissolved chemicals. (This avoids combining calcium nitrate and magnesium sulfate in concentration, which will precipitate calcium sulfate). A twin-head proportioner is thus required.

The rate of injection by the proportioner heads determines the necessary concentration of the nutrient solutions. For example, if each head injects 1 l of stock solution into each 200 l of water passing through the irrigation system, the stock solutions must be 200 times the concentration desired.

The total dissolved salts of the nutrient solution/water actually delivered to the plants must be checked; the proportioner pumps must be examined separately to verify that the pumps are operating correctly. Merely testing for total salts at the end of the system is not sufficient; if one head is ejecting too much solution, and the other too little, total salts may appear to be appropriate, but the ratio of elements may be badly skewed.

9.5 EC Controller (Conductivity)

An EC controller provides the best method to control the salt concentration (salinity) by measuring and controlling the electrical conductivity (EC) of the solution. The EC controller monitors and displays the conductivity of the nutrient solution; it also operates the metering pumps when the measured conductivity falls below a preset value until the optimum value is restored. Electrical conductivity is given in either microsiemens/cm (μS/cm), millisiemens/cm (mS/cm) or micromho/cm (μmho/cm). Two main types of conductivity cells are available: a dip cell suspended in the solution (suitable for small installations), and a flow-through type incorporated in the pipeline. The EC of a solution increases by about 2% for every degree Celsius increase in temperature, so it is important that the conductivity controller is equipped with automatic temperature compensation (this is standard option in most conductivity controllers). It is difficult to provide a general recommendation for the optimum conductivity setting, since it will vary according to the crop grown, the season, the stage of growth, and the quality of the water. A balanced nutrient solution suitable for the growth of most plants has a conductivity of about 1500 μS/cm (Holder and Christensen 1988).

9.5.1 Electrical Conductivity Measurement and Interpretation

Definition: conductivity is a measure of the electrical resistance of water, nutrient solution, or effluent from a growing bed or container, used to determine the content of ions in solution to describe levels from deficiency to excess. Electrical conductivity is determined using a standard conductance cell and salt bridge, the reading espressed as:

$$\text{Specific Conductance (L, mmhos/cm)} = \frac{\text{cell constant (0)}}{\text{resistance (ohms)}}$$

The common methods of expression for L are:

millimhos (mmhos) –mhos $\times$ 10^{-3}/cm
* –mhos $\times$ 10^{-5}/cm
micromhos (μmhos) –mhos $\times$ 10^{-6}/cm

*this method of expression is used to give whole numbers rather than decimals when expressed as mmhos.
(Example: 0.1 mmhos = 10 mhos $\times$ 10^{-5})

Conversions
Osmotic Pressure (atm) $= 0.28 - 0.36\ L_{mmhos/cm}$
Milliequivalents of salt-liter $= 12.5\ L_{lmmhos/cm}$
Parts per million of salts $= 640\ L_{mmhos/cm}$

The electrical conductivity of the nutrient solution is in millisiemens (mS/cm^{-1}) usually within the range 2–4 mS cm^{-1} tomato and cucumber at 2.5–3.0 and lettuce at 2.0 mS/cm^{-1}.

Suitable conductivity levels will be somewhat higher in "hard water" (high calcium) areas than in areas with "soft" (low calcium) water.

9.6 CO_2 Controllers

CO_2 controllers act as a thermostat does for temperature. When the CO_2 level drops below the setpoint, the contact comes on, and the the level exceeds the setpoint by a certain amount, the contact is shut off. One controller can be used for several compartments if a scanner or multiplexer is used.

The contacts are mainly solenoid valves for liquid gas and burner ignitions for furnaces.

9.7 DO Controller

An oxygen-sensitive membrane electrode composed of two solid metal electrodes in contact with a supporting electrolyte is the dissolved oxygen

Fig. 9.5. An example of a CO_2 controller

(DO) measuring unit. A membrane selective to molecular oxygen separates the measuring unit from the test-nutrient solution.

The electric diffusion current is linearly proportional to the concentration of molecular oxygen. The current is converted in the measuring apparatus to concentration units as milligrams per liter.

The membrane permeability and the electrode sensitivity are temperature-dependent. Most DO measuring apparatuses include nutrient solution temperature measuring units (thermistors) and calculation tables. DO measuring units can be installed to activate aeration or irrigation apparatuses when the DO falls below a specified level.

10 Culture and Cultivation

There are small differences among the optimal nutrient compositions for various crop species. No less important are the differences associated with the development stages of the plant, the seedling-, vegetative-, flowering-, and fruiting-stages.

For commercial purposes the harvest stage of the plant, rather than the esthetically pleasing stage, is essential to obtain the maximum yield when the market price is best. Increasing yield through addition of CO_2 or light will have the greatest effect when the nutrition is balanced. Every item has its price, and it is difficult to give general advice in this area. Consultation with scientific experts could improve the income from the investment.

Manipulation of plant development or decreasing plant growth time according to market requirements is still at the beginning stages, and soilless culture methods may provide one of the keys to these concepts.

Many recommendations and information provided in this book are based on farmers' observations and are not always applicable in individual cases. Experience is often the best teacher.

Plant cultivation and culture comprise those operations the gardener performs in changing or controlling the plant environment to obtain maximum yields. Hydroponics is a distinct type of plant culture, in many respects similar to soil culture; however, the radical differences in the growth medium give rise to certain practices and problems of cultivation that do not occur in or are different from those in soil culture.

The information on cultivation given in this book cannot supplant a sound understanding of general gardening techniques.

10.1 Plants Adaptable to Hydroponic Culture

All plants may be grown hydroponically. Whether crops are economic or not will depend on location and other factors. Most root crops are not economic, while plants with large aerial systems, e.g., tomatoes, peas, and cucumbers, are more economic, since their relatively small root systems will result in large yields. Plants supported aboveground are generally more economic, but nonclimbing plants, such as lettuce and peppers, may be equally so under certain circumstances. Flowers are generally more economic because superior quality is obtainable. Also interrow crops like lettuce may be grown with some other crop (Bently 1959; Jensen 1980; Morgan and Tan 1983; Resh 1983; Ministry, Ontario 1988; Mortley et al. 1991).

10.1.1 Plant Variety

The same varieties used in soil culture are usually successful in hydroponics. Other varieties may be even more economically worthwhile because of the conditions peculiar to hydroponics, e.g., absence of certain root diseases and use of supports.

10.1.2 Planting Seeds

In all surface irrigating methods, regular soil seeding and young plant care procedures are used. In subsurface watering and methods, the following procedures are common.

Before seeding, the bed should be wet. Small seeds, such as tomato, should be sown 2–3 cm deep, larger seeds, such as cucumber, 4–6 cm deep. Very small seeds, such as lettuce, should be sown immediately below the first media layer. Some seeds, such as melon and cucumber, are best soaked for 24 h and sown when still wet.

10.1.3 Watering Seeds

Seeds should be kept moist but never flooded. Moistness may be maintained by irrigating once or twice daily and covering the growth media until the first leaves appear, or by irrigating three or four times daily with no covering. Small seeds need more frequent irrigation than larger ones.

10.1.4 Care of Young Plants (in Subirrigating Methods)

When first leaves develop, a diluted nutrient solution is used; since the root should be kept wet during the first 2 weeks, more frequent irrigation is necessary. The root crown, however, should not be kept wet. A dilute solution may be left in the beds during the day, for the first 2–3 days. If sunlight is strong, young plants must be shaded. Thinning out should be done fairly early to assure development of strong plants.

10.1.5 Transplanting

Transplanting seedlings without attached soil balls results in destruction of the side roots and especially of the rootlets. After transplanting, seedling transpiration area is not in proportion to the water, and nutrient uptake area and the ratio between root and shoot is changed often, resulting in temporary wilting. The development of new roots and rootlets restores the root-shoot ratio.

These difficulties are usually not encountered when transplanting within the framework of any soilless culture methods, in which there is a minimal amount of root destruction. Most soilless-grown plants either do not develop any root hairs at all, or only microscopic enlargements.

When transplanting seedlings from soil to soilless conditions, a quite common situation, the same problems arise as under soil-to-soil circumstances. However, the "recovery" of the seedlings in soilless culture is quicker than in soil.

Any attempt to promote new and larger root formation merely accentuates the setback suffered during transplanting. The factors affecting root formation are both internal, such as the amount of metabolites available and their translocation, level of plant hormones, etc., and external, such as climate, humidity, soil, water, pH, etc.

Some soilless growers claim that certain macronutrients influence root development in seedling growth stage. According to some experiments, no significant differences are observed if high levels of NH_4, Mg, or K are used.

While transplanting soilless-grown seedlings with attached growth media balls, as under rockwool conditions, no special treatment is required.

Seedlings from soilless conditions, without balls, may be directly transferred or transplanted, but previously hardened plants will develop better. Vegetable and flower seedlings are generally transplanted 6–12 weeks after sowing: they should be hardened pluming the last week by graduated raising of the solution's osmotic pressure by irrigation with a more concentrated nutrient solution. Shading is removed during hardening. Hardened plants absorb water immediately after transplanting if irrigated with a dilute solution or tap water. Roots develop quickly and the plants continue to grow well.

Freshly transplanted seedlings should be irrigated more frequently for 2–3 days.

When transplanting from soil to soilless conditions, the roots of the soil-grown seedlings should be washed before transplanting. The best way is by immersing the seedlings in standing water for a few minutes, which will dissolve the soil particles without greatly damaging the main roots. Disinfecting agents may be added to the water to prevent introduction of soil-borne diseases. Transplanting should follow immediately after washing, and the irrigating of the new plants should not be delayed.

10.2 Tomato

10.2.1 Plant Spacing

For most varieties, row spacing of 40–60 cm with 25 cm between plants is recommended. Both two- and three-row planting give the same yields in 1-m beds. Paired rows are recommended for wide beds, spaced 30–35 cm apart with at least 50 cm between each pair.

Side shoots should be pinched off as soon as they appear, since they waste the plant's energy. Side shoots develop in leaf axels and fruit trusses between leaf bases, and can be easily distinguished. The main stem is tied to the support system; it will produce the best fruits. The plants should be hitched to a wire around which the plant is wound; winding should also be done once a week. If the main stem reaches the top of the wire (about 2 m), it will droop. In many varieties this is the best time to cut the terminal growing point, especially since most fruits develop in the first 2 m.

10.2.2 Cold Treatment for Seedlings

Cold treatment subjects young tomato seedlings, just after the seed leaves (cotyledons) unfold and the first true leaves start to appear, to 10 to 13 °C (night/day) air temperature for approximately 2 weeks, while providing as much light as possible for 9 to 12 h a day. Low plant temperatures at this stage of growth result in a small number of leaves below the first cluster (and therefore earlier flowering), whereas low root temperatures result in branched clusters (many flowers in the first and possibly the second cluster). The cold treatment increases the number of flowers, but does not influence the setting of fruit. If later conditions for fruit setting are right, then a greater number of flowers will set due to the increased number of blossoms. If, however, the temperature for fruit set is too low, the pollen will not germinate and grow normally, which will result in poor fruit set and cat-faced fruit. When the cold treatment is to be used, seeding should take place 10 to 14 days earlier than usual to compensate for slow growth during the cold treatment. The growth medium in the seedling trays must be sterile, because when plants are grown at a relatively low temperature, the danger of damping-off is increased. Such treatments have been used in soilless culture methods but not always found superior to regular treatments.

10.2.3 Light for Seedlings

During the first few days after pricking out, it is economical to maintain high light intensity for approximately 16 to 18 h daily. This treatment will result in a shorter propagation time and heavier, stronger, sturdier transplants. There is no advantage in using low-intensity incandescent lighting on tomato plants to extend the daylight period (Tremblay et al. 1984).

10.2.4 Temperature Control for Seedlings and Fruit Production

Temperature control recommended by many growers is summarized in Table 10.1.

Root temperature can be easily controlled in many soilless culture installations. Reducing the night air temperature lowers fuel costs but also delays fruit picking by 2–4 days for each degree Celsius that the air temperature is lowered. Despite the reduced early yield obtained from tomato plants grown at low (11–13 °C) night air temperatures compared with normal (15–16 °C) conditions, total yield and value for the whole season are often higher under the former conditions (Hurd and Graves 1985; Marsh and Albright 1991).

One feature of systems of water culture is the ease with which root temperature can be manipulated to suit the requirements of individual crops. Root zone warming (RZW) increases the rate of growth which is associated with the production of larger leaves and greater rates of both photosynthesis per unit leaf area and of translocation. The form and structure of the roots are changed, roots from heated solutions being longer and more extensively branched. These changes have no or very little effect on higher yields.

Tomato plants at 13 °C solution temperature were less than half the size of those at 28 °C root temperature, and their leaves showed symptoms of iron, phosphorus, and calcium deficiency.

Table 10.1. Recommended temperatures for Raising tomato transplant

Growth stage	Air temperature in °C		Root temperature in °C
	Day	Night	
Seed germination	24	24	
For 2 weeks after colyledon expansion	13	13	13
After pricking	21	18	18
Plant development	23–25	18	18

A number of cultural problems arise when tomatoes are grown at reduced night air temperatures coupled with solution heating (23–30 °C). As the solution temperature is raised above 20 °C, root systems become more extensive, leading to vegetative growth greater than considered desirable and delayed fruit setting and development. A reduction in fruit quality is often observed owing to "hollow" fruit or to increased incidence of blossom-end rot.

It has been found that solution temperature and air (night) temperature influence fruit production independently, and warm (20–30 °C) solution temperatures do not offset the delaying effects of low air temperature. Whenever low night air temperature (5–12 °C) regimes are present, it is important to maintain solutions at a minimum of 13–15 °C (Marsh et al. 1987; Moorby and Graves 1980; Morgan and Ohaire 1978).

The onset of fruit picking is usually associated with a period of root death. Some root death at this stage is regarded as a normal physiological phenomenon (Papadopoulos and Tiessen 1987).

10.2.5 Disorders in Plants and Fruits

Temporary wilting of plants after transplanting often occurs, mainly due to a lack of water-nutrient solution (infrequent irrigation) and/or insufficient root surface for moisture absorption (root:shoot ratio) and/or too many leaves (transpiration area) compared to roots (absorption area). Reduction of shoot transpiration (shading or reduction of temperature) will overcome this period of wilting.

Fruit cracking may occur in small, newly set fruit or when the fruit is at or approaching the ripening stage. Growth cracks or splits that appear in small developing fruit, usually at the blossom end, seem to be associated with an uneven growth rate and also with excessive humidity and cool night temperature. Any factors that can be regulated so as to encourage steady growth reduce the danger of its incidence.

In mature fruit, cracks usually occur on large fruit during bright, hot weather, and their occurrence is increased by high salt content of the medium. Cracking often happens shortly after the plant has been topped, especially when the lateral shoots have been kept pruned

Rough fruit occurs most often on the early clusters and is usually caused by low temperatures during the development of early flower buds. Another cause is transplanting before the first flower buds can be seen on the plants, and where the transplanting is followed by dull weather.

Excess nitrogen available to the plant after transplanting may result in rapid growth rates and in curling of new leaves, and thus in production of rough fruit. Increased potassium levels and decreased nitrogen levels during the growing season may help prevent cracking.

Excessively high or low temperatures, too high moisture, or an extremely unbalanced nutrient solution may cause puffiness ("pockets")

in tomatoes, if any of these conditions arise when the fruits have reached about two-thirds of their size: the fruit wall develops normally, but the inner tissues are retarded. Fruits are lightweight and only partially filled. Puffiness does not develop with a normal nutrient supply, provided extremes of temperature or moisture are not present.

Leaf roll may often be associated with insect attack or virus diseases. Excessive moisture or very strong sunlight may cause leaves to roll in a similar fashion. Overlapping is usually attributed to an excess of accumulated organic reserves in the leaves; however, if leaf roll is detected, deficiency or excess or nutrients should be tested for. Different varieties vary in their susceptibility to leaf roll.

The use of saline waters may cause a water-soaked, brown-black spot to develop on the fruits, opposite the calyx (blossom-end rot). This may reduce yields by 25–35%. Blossom-end rot caused by salinity is similar in appearance to that caused by calcium deficiency. Some varieties (e.g., Merrimande) are more sensitive to blossom-end rot caused by salinity, while others are not (e.g., Moneymaker).

Blossom-end rot may also develop if the solution's nitrogen content is high and the calcium content low, apparently because of inhibition of calcium uptake.

A low phosphorus content may lead to uneven ripening. Flower drop is mainly caused by high temperature, strong wind, heavy rain, hail, or insects. Flowers are not fertilized at high temperatures.

10.3 Cucumber

The cucumber is a semi-tropical plant which grows best under conditions of high temperature, humidity, light intensity, moisture supply, and nutrient level. If these conditions prevail, and insects and diseases are controlled, growth is amazingly rapid and a large volume of fruit can be harvested. However, cucumbers are highly sensitive to unfavorable conditions. Each growth factor must be carefully controlled because levels varying only slightly from the optimum can cause marked reductions in growth and yield. Most cucumbers now grown in greenhouses are the long seedless type.

When using standard cucumber varieties seeds are soaked in water for 24 h before sowing at a depth of 3–4 cm, in rows 50 cm apart, with 15–20-cm spaces between seeds in the row.

Until germination, irrigation practice is the same as for tomatoes. If saline water is used, the seeds should be germinated either in tap or less saline water, or germination will be delayed.

Cucumbers may be transplanted without damage to the side roots if the seedlings, in individual balls, are directly transferred to the beds. They are supported like tomatoes. Well-supported cucumber plants are more easily cultivated and harvested. They should be irrigated about 30% more frequently than tomatoes. Late fruits are usually smaller and more rounded; this cannot be prevented by altering the nutrient solution.

Male and female flowers develop on the same stem, so that self-pollination is the rule. In most varieties, the first five to eight flowers on the main stem are male, these being followed by alternating female and male flowers. However, the further the flower from the roots, the greater the frequency of female flowers. The first flower on the side stems is usually female, and is followed by alternating male and female flowers. Side stems generally have more female flowers than the main stem, but if the latter is cut early enough, one side stem will have a flower pattern to the main stem.

Cucumber flowers pass through a bisexual stage and after a certain point become either male or female. Sex expression in cucumbers is affected by growth hormones, temperature, and day length. With short, low-temperature days, the first female flower will appear earlier and at a lower node on the main stem.

Changes in inorganic nutrients, especially of nitrogen, affect sex expression. The addition of ammonium sulfate, increasing the nitrogen rate by 50% in the winter, may change male:female flower ratios from 1:8 to 1:4. This change is especially marked in saline conditions. "Large-fruit" cucumbers usually have a female flower at every node.

Seedless cucumbers are 28 to 50 cm long, uniformly green and bitter-free, with a thin, edible skin. This cucumber is parthenocarpic, which means it will set fruit without pollination, If pollination occurs seed will form and fruit becomes clubbed at the blossom end and develops a bitter taste. To prevent pollination, bees must be kept out of the greenhouse, and only gynoecious cultivars (all female flowers) should be grown. If a few male flowers appear (as may happen with some cultivars that are not 100% female), they should be removed as soon as possible. Nutrient solution balance may influence sex expression.

Cucumber seed requires moisture, oxygen, and heat for germination. Seed should be grown in a moist and well-aerated medium, such as a good soilless mix prepared for this purpose.

The optimum germination temperature of cucumber seed is 27–28 °C.

Following seeding, the seed is covered with about 1 cm of medium which must be kept moist. This is best done by covering the flats or seedbed with plastic or by placing a shallow "tent" over them. On sunny days, a reflective cover may be needed to prevent excessive temperatures under the cover.

Four days after seeding, when the cotyledons have expanded, the seedlings must be transplanted into 10- to 12-cm pots. For the first few days after transplanting, the temperature should be maintained at 23–24 °C. After the plants are established, the night temperature can be lowered a few degrees, but it should never drop below 20 °C. Minimum day temperature is 23 to 25 °C, depending on the light intensity. Temperatures higher than 28 °C should be avoided. As the pot temperature should be at least 26 °C, bottom heat may be necessary.

Initially, the pots with seedlings can be placed close together. After 10 to 14 days, when the leaves begin to touch each other, the plants should be spaced farther apart to expose them to more light and avoid undue stretching. Supporting the plants will keep them growing straight, prevent breakage during transport, and makes stringing easier after they are planted in the greenhouse.

The optimum plant population is considered to be 120 to 160 plants per 100 m^2. A number of factors, such as cultivar, growing season, and media, are to be considered when deciding on the plant population. Some cultivars are more vigorous than others and should, therefore, be spaced farther apart. The growing medium affects vegetative growth. Substrate-grown (e.g., rockwool) crops are somewhat less vigorous. When a crop is expected to be less vigorous, the plant population should be increased.

The rate of development, length of fruit, color, and balance between vegetative growth and fruit development of seedless cucumbers are all affected by temperature.

The temperature regime for the first 6 to 8 weeks after planting out should be a minimum temperature of 20 °C at night and a daytime temperature of about 23–25 °C. Temperatures higher than 30 °C should be avoided. After the initial period and when the plants have started to produce, the night temperature can be gradually reduced to as low as 17 °C. However, the day temperature should be maintained at 23 or 24 °C so that the 24-h average temperature is around 20–21 °C.

To prevent condensation on the plants, and particularly on the fruit, it is very important when low night temperatures are maintained that the daytime temperature be reached by daybreak. This rise in temperature must be slow and gradual and may have to start as early as 2 h before sunrise.

When ventilation is required to reduce the temperature, care must be taken that the relative humidity does not drop too quickly. A high temperature in the greenhouse is not as dangerous as a sudden large reduction in humidity.

The root-zone minimum temperature of 19–20 °C is required, while 22–2 °C is optimum (Sonneveld 1980; Adamson and Maas 1981; Ministry of Agriculture and Food, Ontario 1988; Adams et al. 1989).

10.4 Lettuce

Lettuce is easy to grow, and no precautions are needed. It may be seeded or grown from transplants. Leaf lettuce is planted at 15 × 20 cm, head lettuce at 20×20 cm. The media must be kept moist from time of sowing to a few days after germination.

If crops are to be continuous, the bed may even be sown between the rows about 10 days before harvest. The media will then be moist enough for germination of fresh seed, due to shading provided by the mature plants. Growing space may thus be economized.

Lettuce is also a good inter-row crop for tomatoes etc.

High root temperatures (up to 20 °C) increase only the early growth. Most lettuce varieties can be grown in saline conditions, but taste is affected. Some varieties are salt-tolerant and even taste better when grown under saline conditions. Even when using highly concentrated nutrient solution, under saline conditions many varieties of plants become progressively smaller and "harder" in appearance, with leathery leaves (Morgan, Tav 1983; Van Os, Kuiken 1984).

10.5 Flowers

10.5.1 Carnation

Planting is as in soil, with either 20×20 or 15×20 cm spacing; 30–40 plants per m^2 are recommended for in-season crops, 50 per m^2 for out-of-season crops. Carnations are less frequently irrigated than tomatoes or other vegetable crops.

It was observed that "thirsty" plants produce better flowers, while excess water causes the sepals to split apart, as do high temperatures.

The nutrient solution is more concentrated. Young seedings need about 50% more nitrogen than supplied by the ordinary nutrient solution. In the vegetative stage, and especially during flower formation, the phosphorus and potassium levels should be raised by 30–40%. Excessive nitrogen during the vegetative stage will cause long, weak stems.

Under saline conditions, stems are generally 10–15% shorter, but they are thicker and the flowers stronger.

10.5.2 Gerbera

Plants should be spaced at 30×30 cm. Double amounts of the microelements are recommended (especially iron, copper, and manganese).

10.5.3 Chrysanthemum

A solution temperature of 20 °C for the first 3 weeks of the vegetative phase will give strong plants. Solution heating (20 °C) combined with low air temperature increases shoot fresh weight and improves quality, particularly bloom size.

During rooting, a low-strength nutrient solution is recommended. During growth, normal strength is required. Double-strength solutions or low saline condition results are similar to those for normal strength.

10.5.4 Orchids

There are certain indications that the most outstanding feature in soilless culture will be the growing of high value crops such as certain species of orchids.

Even though most orchids are typical epiphytes, and according to the literature can be grown "only" in organic growth media, and often only a limited organic growth material is recommended, there are strong indications that many of these varieties may be grown successfully (and maybe even with better results) under soilless culture methods.

Orchid flowers in soilless culture still require basic investigation, but some knowledge is already available (Penningsfeld 1980a, b).

Seedlings of *Dendrobium cattleya* and *Phalaenopsis* develop in gravel culture. Transplanting *D. cattleya* and *Phalaenopsis* from 'soil culture methods is difficult. Some growers have complained that a reduction of flowers was observed (*Cattleya* and *Oncidium*), while other growers did not find any difficulty.

Higher and earlier yields than under normal conditions are claimed for *Cattleya*, *Phalaenopsis*, *Paphiopedium*, but not for *Cymbidium* and similar types, and perhaps this is true for all pseudo-bulb varieties.

The suggested nutrient solution (by Penningsfeld 1980a, b) is (in 10 l) calcium nitrate 100 g ammonium sulfate 50 g; 25 g each monopotassium phosphate, magnesium sulfate, and ferrous sulfate, and 1 cc of stock solution having in 1 l (in g):

Boric acid	2.86
Manganese chloride	1.81
Zinc sulfate	0.22
Copper sulfate	0.08

11 Establishing a Commercial Soilless Culture Unit

When considering the establishment of a commercial soilless-culture unit, date or tests are required for determination of geography and climate conditions including yearly charts of:

- day length,
- light-energy average, minimum and maximum,
- temperature average, minimum and maximum,
- precipitation type and amount,
- windspeed,
- humidity,
- possibility of extraordinary weather conditions – hail, heavy storms, etc.

All this information is available at local weather stations and helps in the preparation of precautions for overcoming extreme situations, in addition to calculation of the most suitable crops to be grown according to local and export prospects.

For the soilless-culture unit it is also important to have information on the water and growth media to be used:

A. *Water*

1. Amount available and variables during the year.
2. Water analysis and variables expected during the year.

The water analysis should include EC (salinity); pH;
Cations: Ca, Mg, Na, K;
Anions: CO_3, HCO_3, SO_4, Cl, NO_3.

Of course, if undesirable ions such as boron are suspected in the water, this amount should be determined and taken into consideration when preparing the nutrient solution.

B. *Growth media:* Gravel, sand, volcanic ash, river mix, etc.

The following quick tests will give preliminary indications on the quality of the growth media

Pour about 10 cc of a 5% sulfuric acid of hydrochloric acid solution into a glass tube and add two teaspoons of the media, if very few or no bubbles are released over a period of a few minutes, the media is probably suitable for plant growth. A final decision can be made only after qualified laboratory testing.

The best applied test is, of course, growing plants on a soilless-culture pilot plant scale for one to two seasons and comparing the growth with that of established growth media.

Higher capital and operating costs help to explain why many types of soilless culture techniques are not more widely used. Most applications are either limited to areas where other techniques are not feasible or used for crops having a high cash value. Soilless culture today is essentially limited to controlled environmental systems, rather than outdoor growing. Increased use of soilless growing basically hinges on the development of systems that will make the technique economically competitive with other growing systems.

As already mentioned, the main sytems recently found to be economical and promising in the near future are:

1. Sand culture method, especially in sandy and dune areas.
2. Rockwool culture, and foam plastic culture.
3. Bag culture.

and under special conditions

4. Gravel culture in restricted areas.
5. NFT, only if better means of disease and oxidation control are found.
6. Aeroponic – only if initial investment could be reduced tremendously.

11.1 Economic Considerations

Since there are so many variable factors involved, one hesitates to print these data and to assume responsibility for them. It is often difficult to give a quick, comprehensive answer to questions such as: "How much is produced in 1 m^2 soil?" or "How much water is required per m^2 in both conditions?" The questions should really be: "What is the investment and turnover capital required to produce a certain yield in both conditions?" and "What is the amount of water required to produce, for example, a ton of tomatoes under both conditions?" (Schwarz 1968; Glaub and Trezek 1981; Muller et al. 1988; Otten et al. 1989; Marsh and Albright 1991).

The costs of soilless-culture systems, excluding the protective structures in which they will normally be housed, are given in Tables 11.1 and 11.2.

The costs quoted in this section assume that all necessary services, e.g., a reliable electricity supply and piped water of adequate purity, are already available on site. These costs would be considerably higher in distant lands with high shipping charges, and do not include the spare pumps, electrodes, etc. which it would be necessary to have available where prompt repair or replacement services were unavailable.

Table 11.1. Capital costs for 1 ha of some soilless-culture systems. (After Collins and Jensen 1983)

Bag culture	\$38 000[a]
Rockwool culture	\$55 000[a]
NFT	\$81 000[a]
Sand culture	\$50 000[b]

[a]Excluding construction costs.
[b]Including construction costs

Table 11.2. Capital costs per ha of some soilless-culture systems. (After Van Os 1982)

	\$/ha
Rockwool culture	37 800
Rockwool (stationary water)	40 800
Peat bags	26 500
NFT	56 600

Table 11.3. Cost comparison between troughs filled with sand versus sand over the entire greenhouse floor (1000 m^2)

Material for beds	Troughs (above-ground) filled with sand (US \$)	Sand over entire floor (US \$)
Wood or concrete – 3280 m	9000	–
Construction – labor	3000	–
Drain pipe – 3.1 cm (1 1/4 in) PVC pipe \$0.46/m	1500	1500
Plastic liner – 2 layers 6 mm each \$0.22/m^2	1742	1742
Sand – \$5.26/m^3 – 608 m^3	3200	
1368 m^3		6400
Placement of sand – labor	2000	2000
Total	18942	11642

If the surface of the bed is level, the tube should not be over 15 m long. If the surface of the bed is sloped 15 cm the length of tube can be 30 m. The supply manifold must be at the high end, where the water is fed into the tubes. On a level bed, the supply manifold may run across the center of a 30 m bed to supply water to 15 m lines off either side

The current cost of the components of 1 ha of an NFT system in the UK, including a fully adjustable metal support system, is about 45 000 pounds. The most expensive item in this estimate is the galvanized iron support system (trays and stands), accounting for 58% of the total cost. Thus, systems laid on expanded polystyrene on a sloped soil surface have a far lower capital cost. A cost comparison between the two main sand culture methods are described in Table 11.3.

Table 11.4. Estimated cost of various commercial hydroponic systems

Rockwool culture	$65 000
Sand culture	
– sand dunes	35 000
– sand beds (plastic)	65 000
– using gravel scoria (volcanic tuff) including 150 km gravel transportation	65 000
Bag culture	
– using scoria gravel including 50 km gravel transportation	65 000

In Israel, the costs are estimated as in Table 11.4 (per ha) (1991).

11.2 Increased Technical Demands on the Management

It would be difficult to improve on the assessment of this topic given by Stoughton (1969) (with some major changes):

Much harm has been done to the proper appreciation of the value of the systems by over-enthusiastic writers who imply and often state categorically that soilless culture is easy and requires little knowledge. This is not true; to make a real success of the methods, it is necessary to have some knowledge of plant physiological principles and even of elementary chemistry, as well as a sound knowledge of how to grow a crop. It should be clear that soilless methods only provide a greater degree of control, and if knowledge of how to use that control is lacking, failure is inevitable.

No one should embark upon serious application of these methods without first studying the subject extensively and, if in doubt of his own knowledge of the scientific principles involved, ensuring access to expert advice at short notice.

This having been said, it is clear that the great increase in soilless culture during the last decade, the use of sand culture, bag culture, rockwool culture, and NFT, coincides with the development of great expertise amongst research scientists, specialist advisers, and the growers themselves. Soilless-culture techniques are not an easier way to produce good crops, they are demanding techniques by which appropriately skilled staff can, under suitable conditions, produce high quality crops, the higher capital and operating costs being offset by high yields and savings in labor. Successful use of hydroponic techniques requires not so much a traditional "grower" but rather a forward-looking professional prepared to accept the technical challenge. The new manager has to be or to become an applied scientist with a very wide range of skills.

12 Some Future Aspects

12.1 Plant Nutrients

No discovery of a new essential element for improved plant growth is expected. The list is complete; but what is it about the 16 known essential elements that make them a requirement for most plants? There is no clear answer. Comparison of their location on the ion element tables offers no reason for their being essential. Related chemical elements could have a similar effect on the plant nutrient mix! Why is potassium so dominant in plants and the "related" sodium so dominant in animals? Why is the magnesium ion dominant in the chloroplast and not "relative" like iron?

It was previously mentioned in the discussion of plant nutrients that some nonrequired elements can partly substitute for a "related" essential element, when the essential element is deficient (bromine can sometimes be substituted for chlorine in this respect). Further research in this field will have no practical effect on applied soilless culture.

The essential chemicals are relatively inexpensive and plentiful. The situation will be altered, however, should there be any tremendous change in plant development by adding or changing an ion. So far, no fair hypothesis has been presented in this connection, but a breakthrough could occur.

The ideal nutrient solution for any plant and growth condition is still quite wide-ranging, and developments in this field are not expected. The ability to adjust nutrient solution composition for a certain plant growth stage (such as vegetative, flowering, fruit setting) is still an open field. The closer we come to optimal environmental growth conditions, the more influence and practical value this adjustment and others (such as adjustment of CO_2, pH, and nutrient solution concentration) will have on plant growth. In the final stages of growth, some plants may require fewer or no nutrients at all. This field of research beckons.

12.2 Initial Investment

Reducing the initial investment costs for an installation will no doubt be the main goal in improving the economics of soilless culture crop development. The second goal will be reduction of the running costs of the installation. The majority of "discoveries" in recent years have been in these directions.

The "new systems" primarily comprise improvements in growth media, containers, and irrigation systems, and continued developments in this direction, combined with improvements in environmental protection such as construction of greenhouses, glass/plastic coverings, and efficient heating/cooling systems, are expected to increase the commercial value of soilless-culture crop development. The percentage of the investment cost of soilless-culture units is relatively low compared to that of environmental control under closed growth conditions. Therefore developments in less expensive soilless-culture growth, where the soilless-culture unit investment is relatively high, will be more cost-effective than under open or semi-closed growth conditions (Wilcox 1982).

12.3 Growth Media

The application of rockwool to plant growth has been one of the most important recent developments. The environmental problem of what to do with the used rockwool still exists. The future will witness production of similar growth media that are environment-friendly.

Production of foam plastic (even on-site production of foam plastic) will have an important effect on increasing yield. Invention of a system composed of expanding plastic inside plastic tubes will permit insertion of seed strips as well as thin plastic drippers directly into the tubes, to provide a handy automatic growth system for both open and semiclosed conditions.

12.4 Advanced Sand-Culture Methods

These will extend to dunes especially in coastal areas, using saline water sources for the prospective higher salt-tolerant culture species. Inexpensive sea-water desalination techniques will result in similar developments.

Important strides in increasing plant longevity or decreasing growth time until harvest, and in reducing waste parts in plants to free energy for production of required plant products, especially if effected by nutrition or root environmental control, will have great influence on soilless-culture economics.

There is a general tendency in plant production research to reduce the dependence on climatic and environmental factors through such means as selecting or breeding of species with broader growth ranges and greater tolerance to the stresses of temperature, light, salinity, water, and humidity. Developments in this area will also affect plant growth under soilless conditions. Control of the root:shoot ratio within the plant structure itself, and improvement of harvest index ratio through nutrition and root environment condition control will be more pronounced and have greater potential in soilless conditions than in soil.

The challenge of the future will be the remarkable improvement of present soilless-culture methods with a view towards the new fields of Water Culture, Nutrient Film Techniques, Aeroponics and Vertical Growth.

Acknowldgments. The cooperation of all who provided photographs or diagrams for this publication is gratefully acknowledged.

References

Adams P (1980) Nutrient uptake by cucumbers from recirculating solutions. Acta Hortic 98:119–126

Adams P, Graves CJ, Winsor GW (1989) Some responses of cucumber, grown in beds of peat, to micronutrients and pH. J Hortic Sci 64:293–299

Adamson RM, Maas EF (1971) Sawdust and other soil substitutes and amendments in greenhouse tomato production. Hortic Sci 6(4):397–399

Adamson RM, Maas EF (1976) Amount and kind of growth media in soilless greenhouse tomato production. Hortic Sci 11(3):212–213

Adamson RM, Maas EF (1981) Soilless culture of seedless greenhouse cucumbers and sequence cropping. Agric Can Publ 1725/E, 21 pp

Alberry WS, Haggett BD, Svanberg LR (1985) The development of sensors for Hydroponics, Biosensors 1: 369–92

Alexander A (1986) Foliar fertilization. Proc 1st Int Symp on Foliar Fertilization. Martinus Nijhoff, Dordrecht, pp 3–488

Alwan AH, Newton P (1984) Dissolved oxygen root growth, nutrient uptake and yields of tomatoes. ISOSC Proc 5th Int Congr Soilless Culture PO Box 52, Wageningen, pp 81–112

Bauerle WL (1984) Bag culture production of greenhouse tomatoes. Ohio Agric Res Center, Circ 108, 7 pp

Benoit F, Ceustermans N (1986) Growth control of tomatoes and cucumbers in NFT by means of rockwool and poly-urethane blocks. Soilless culture 2:2, 3–10

Bently M (1959) Commercial Hydroponics (2nd edn). Benton Jones J (1983) A guide for the hydroponic and soilless culture grower. Timber Press, Portland, 124 pp, Johannesburg, 750 pp

Berry WL, Wallace A, Lunt OR (1977) Recycling municipal wastewater for hydroponic culture. Hortic Sci 12:186

Blom TJ, Straver WA (1988) Two years of experience with growing roses on mineral wool. ISOSC Proc 7th Int Congr Soilless Culture PO Box 52, Wageningen, pp 103–115

Blom TJ, Ingratta FJ, Hughes J (1990) Greenhouses. Ministry of Agriculture and Food, Ontario, Canada, Publ 40

Boodley JW (1984) Foam substrate applications in North America. ISOSC Proc 6th Int Congr Soilless Culture, PO Box 52, Wageningen, pp 81–112

Butler JE, Loveridge RF, Bone DA (1988) The use of hydroponics for sewage treatment in temperate and tropical areas. ISOSC Proc 7th Int Congr Soilless Culture, PO Box 52, Wageningen, pp 103–116

Carpenter TD (1982) Analysing and managing nutrition of vegetables grown in upright polyethylene bags. J Plant Nutr 5(8):1083–1089

Clarkson DT, Hanson J (1980) Mineral nutrition. Annu Rev Plant Physiol 239–289

Collins WL, Jensen MH (1983) Hydroponics: a 1983 technology overview. Environmental Research Laboratory University of Arizona, Tucson, 119 pp

Cooper AS (1975) Crop production in recirculating nutrient solution. Sci Hort 3:251–258

Cooper A (1979) The ABC of NFT. Grower Books, London, 181 pp

Cotter DJ (1979) Effect of fresh bark media on phosphorus status of tomato plants. Commun Soil Sci Plant Anal 10(11):1381-1384

Crum SJH, Runia WT, Leistra M, Smelt JH (1985) Behaviour of some pesticides in a nutrient film and in a rockwool system. Sci Hort 25:1-9

Davtyan GS (1980) Classification of hydroponic methods of plant production. ISOSC Proc 5th Int Congr Soilless Culture, PO Box 52, Wageningen, pp 45-52

Donnan RS (1988) Carnation cropping in rockwool. ISOSC Proc 7th Int Congr Soilless Culture, PO Box 52, Wageningen, pp 117-134

Ehret DL, Ho LC (1986) The effects of salinity on dry matter partitioning and fruit growth in tomatoes grown in nutrient film culture. J Hortic Sci 61:361-367

Ellis C, Swaney MW (1947) Soilless growth of plants (2nd edn, revised by Eastwood T). Reinhold, New York, 277 pp

Elstner EF (1982) Oxygen activation and oxygen toxicity. Annu Rev Plant Physiol 33:73-96

Emmert FH (1961) The bearing of ion interactions on tissue analysis results. In: Reuther W (ed) Plant analysis and fertilizer problems. American Institute of Biological Sciences, Washington, pp 231-243

Enoch HZ, Zieslin Y, Biran AH, Schwarz M, Kessler B, Shimshi D (1972) Principles of CO_2 nutrition research. Acta Hortic 32:97-115

Epstein E (1972) Mineral nutrition of plants: Principles and perspectives. John Wiley, New York, pp 39, 62, 63

Evants LT (1975) The physiological basis of crop yield. In: Evans LT (ed) Crop Physiology. Cambridge University Press, Cambridge, pp 327-352

FAO (Food and Agriculture Organization) (1990) Plant Production and Protection Pap 101, Rome

Feigin A, Zwibel M, Rylski I, Zamir N, Levav N (1980) The effect of ammonium/nitrate ratio in the nutrient solution on tomato yield and quality. Acta Hortic 98:149-160

Feldman LJ (1984) Regulation of root development. Annu Rev Plant Physiol 35:223

Fischer P (1984) Stability of various forms of chelated iron in nutrient solutions of different pH values. ISOSC Proc 6th Int Congr Soilless Culture, PO Box 52, Wageningen, pp 225-233

Gericke WF (1937) Hydroponics - crop production in liquid culture media. Science 85:177

Gericke WF (1938) Crop production without soil. Nature (Lond) 141:536-540

Gislerod HR, Kempton RJ (1983) The oxygen content of flowing nutrient solutions used for cucumber and tomato culture. Sci Hortic 20:23-33

Glaub JC, Trezek GJ (1981) Investigation of solar-heated hydroponic greenhouse. American Society of Agricultural Engineers, (ASAE) Pap 81-4040, St. Joseph, MI, 21 pp

Graves CJ (1983) The nutrient film technique. Hortic Rev 5:1-44

Graves CJ, Hurd R (1983) Intermittent solution circulation in the nutrient film technique. Acta Hortic 133:47-52

Green JL (1968) Perlite - advantages and limitations as a growth medium. Colo Flower Grow Bull 214, 6 pp

Hall DA, Hitchon GM, Szmidt RAK (1988) Perlite culture. ISOSC Proc 7th Int Congr Soilless Culture, PO Box 52, Wageningen, pp 177-183

Hanan JJ, Holley WD, Goldsberry KL (1978) Greenhouse management. Springer, Berlin Heidelberg New York

Hanger BC (1979) Hydroponic fundamentals. In: Commercial applications of hydroponics; agriculture note ser No 12 Dept of Agriculture Victoria. Paper No 1, 4 pp

Hanger BC (1982) Rockwool in horticulture - a review. Aust Hortic 80(5) 7-16

Harris DA (1976) A modified drip culture method for the commercial production of tomatoes in vermiculite. ISOSC Proc 4 Int Congr Soilless Culture, pp 85-90

Harris DA (1977) Hydroponics, the gardening without soil. Purnell, Cape Town, 207 pp

Harris DA, Maree PCJ (1984) Growing tomatoes in a plastic greenhouse in uncomposted pine bark. ISOSC Proc 6th Int Congr Soilless Culture, PO Box 52, Wageningen pp 245-254

Hewitt EJ (1966) Sand and water culture methods used in the study of plant nutrition. Commonw Bur Hortic Plant Crops Tech Commun 22 (revised)

Hewitt EJ, Smith TA (1975) Plant mineral nutrition. English University Press, London

Hoagland DR, Arnon DI (1938) The water-culture method for growing plants without soil. Calif Agric Exp Stn Circ 347:39

Holder R, Christensen MH (1988) Effect of EC on the growth, yield and composition of cherry tomatoes grown on rockwool. ISOSC Proc 7th Int Congr Soilless Culture, PO Box 52, Wageningen, pp 213–227

Hurd RG, Graves CJ (1985) Some effects of air and root temperature on the yield and quality of tomatoes. J Hortic Sci 60:359–371

Hurd RG, Price D (1977) Root death and mid-crop wilting of tomatoes in NFT. Horticulture Industry Jan 1977 pp 15–18

Islam AKMS, Edwards DG, Asher CJ (1980) pH optima for crop growth. Results of a flowing solution culture experiment with six species. Plant Soil 54:339–357

Jackson MB, Blackwell PS, Chrimes JR, Sims TV (1984) Poor Aeration in NFT and a means for its improvement. J Hortic Sci 59(3):439–448

Jensen MH (1968) Ring and trough culture for greenhouse tomato production. Tech Rep, Environm Res Lab, Univ of Arizona, Tuscon

Jensen MH (1980) Tomorrow's agriculture today. Am Veg Grower 28(3):16, 18, 19, 62, 63

Jensen MH (1981) New developments in hydroponic systems: descriptions, operating characteristics, evaluations. Proc 2nd Annu Conf Hydroponics Society of America 2, Dublin, CA, pp 1–25

Jensen MH (1984) Hydroculture and new technologies for greenhouse vegetable production. Symp Int sur la Serriculture, 26/27 octobre, Université Laval, Quebec, pp 249–304

Jensen MH, Collins WL (1985) Hydroponic vegetable production. Hortic Rev 7:483–558

Jensen MH, Hicks NG (1973) Exciting future for sand culture. Am Veg Grower (Nov) 33, 34, 72, 74

Johnson H (1980) Hydroponics: a guide to soilless culture systems. Division of Agricultural Sciences, University of California, Leaflet 2947, 15 pp

Johnson IR (1985) A model of partitioning of growth between the shoots and roots of vegetative plants. Ann Bot 55:421–431

Jones JB (1982) Hydroponics: its history and use in plant nutrition studies. J Plant Nutr 5(8):1003–1030

Judd R (1982) Bag culture. Am Veg Grower 30(11):42, 44

Kramer PJ (1981) CO_2 concentration, photosynthesis and dry matter production. BioScience 31(1):29–33

Kramer D (1983) The possible role of transfer cells in the adaption of plants to salinity, Physiol Plant 58:549–555

Krizek DT (1982) Plant response to atmospheric stress caused by waterlogging. In: Christiansen I (ed) Breeding plants for less favorable environments. John Wiley, New York, pp 293–334

Leopold AC, Kriedemann PE (1975) Plant growth and development. McGraw-Hill, New York

Luettge U (1983) Import and export of mineral nutrients in plant roots. In: Inorganic plant nutrition. Encyclopedia of plant physiology, vol 15, Springer, Berlin Heidelberg New York, pp 203–238

Maas EV (1986) Salt tolerance of plants. Appl Agric Res 1:12–26

Maas EF, Adamson RM (1981) Artificial media in horticulture – their formulation and fertilization. Agric Can Publ 1726/E

Maher MJ (1976) Growth and nutrient content of a glasshouse tomato crop grown in peat. Sci Hort 4:23–26

Maree PCJ (1981a) Growing greenhouse cucumbers in fresh sawdust. Dept of Agronomy and Pastures, University of Stellenbosch, 12 pp

Maree PCJ (1981b) Commercial production of tomatoes in vermiculite. Dept of Agronomy and Pastures, University of Stellenbosch, 8 pp

Marschner H (1983) Essential mineral elements. In: Inorganic plant nutrition, Encyclopedia of Plant Physiology, vol 15. Springer, Berlin Heidelberg New York, pp 5–12

Marsh LS, Albright LD (1991) Economically optimum day temperature for greenhouse hydroponic lettuce production. Am Soc Agric Eng 34(2):550–562

Marsh LS, Albright LD, Langhans RW, McCulloch CE (1987) Economically optimum day temperature for greenhouse hydroponic lettuce production. ASAE Pap No 87-4023, St Joseph, MI

Massantini F (1973) L'Aeroponica: realizzazioni practiche studi e ricerche. ISOSC Proc 3rd Int Congr Soilless Culture, PO Box 52, Wageningen, pp 83–97

Massantini F (1976) Floating hydroponics: a new method of soilless culture. ISOSC Proc 4th Int Congr Soilless Culture, PO Box 52, Wageningen, pp 91–98

Massantini F (1985) The light and dark sides of aeroponics. Soilless Cult 1(1):85–96

Meiri A, Plaut Z (1985) Crop production and management under saline conditions. Plant Soil 89:253–271

Meiri A, Shalhevet J (1973) Crop growth under saline conditions. Ecol Stud Anal Synth 5:272–290

Ministry of Agriculture and Food, Ontario (1988) Publ 526, Growing Greenhouse Vegetables, pp 14, 17

Moorby J, Graves CJ (1980) Root and air temperature effects on growth and yield of tomatoes and lettuce. Acta Hortic 98:29–43

Morgan JV, Ohaire B (1978) Heated hydroponic solutions as an energy savings technique. Acta Hortic 76:173–180

Morgan JV, Tan AL (1983) Greenhouse lettuce production at high densities in hydroponics. Acta Hortic 133:39–46

Mortley DG, Bonsi CK, Loretan PA, Morris CE, Hill WA, Ogbuehi CR (1991) Evaluation of sweet potato genotypes for adaptability to hydroponic systems. Crop Sci 31:8

Muller PH, Tiessen S, Khosla L, Otten F, Rey H, Jackson A (1988) Energy-saving hydroponic greenhouse pilot project: economic aspects. Production Analysis Division, Policy Branch, Agriculture Canada, Ottawa

Namioka H (1976) Kuntan as a substrate for soilless culture. ISOSC Proc 4 Int Congr Soilless Culture, pp 289–302

Nir I (1980) Growing plants in aeroponic growth systems. Acta Hortic 99:147

Otten L, Muller PG, Thiessen H, Khosla S, Rey F, Jackson HA (1989) Energy-saving hydroponic greenhouse pilot project. Can Agric Eng 31:147–152

Papadopoulos AP, Tiessen H (1987) Root and air temperature effects on elemental composition of tomato. J Am Soc Hortic Sci 112:988–993

Penningsfeld F (1980a) Growing orchids in expanded clay. ISOSC Proc 5 Int Congr Soilless Culture, pp 313–322

Penningsfeld F (1980b) Orchids in hydroculture (in German). Gartenbau Gartenwiss 80(24):527–30

Pickford RJJ, Hall NJ, Wight PN (1985) The physical and biological properties of rockwool as a growth medium. Poster to the Assoc App Biologist Conf, Univ of Bangor, Wales, March 1985

Porter MA, Grodzinski B (1985) CO_2 enrichment of protected crops. Hortic Rev 7:345–398

Potter R (1977) Peat culture tomatoes. Ministry of Agriculture Fisheries and Food, South Coast Glasshouse Unit, Chichester, 9 pp

Rendig VV, Taylor HM (1989) Principles of soil-plant interrelationships. McGraw-Hill, New York

Resh HM (1983) Hydroponic food production. Woodbridge, Santa Barbara, 335 pp

Robson AD, Pitman MG (1983) Interactions between nutrients in higher plants. In: Inorganic plant nutrition, Encyclopedia of Plant Physiology, vol 15A. Springer, Berlin Heidelberg New York, pp 147–158

Roeber R, Schaller K (1985) Planzenernaehrung im Gartenbau. Eugen Ulmer, Stuttgart

Rogers HT (1983) Greenhouse of the future. Greenhouse Grower 1(1):72, 74, 76

Salisbury FB, Ross CL (1969) Plant physiology. Wadsworth, Belmont, CA

Schwarz M (1968) Guide to commercial hydroponics. Israel Universities Press, Jerusalem, 136 pp

Schwarz M (1972) Influence of root crown temperature on plant development. Plant Soil 37(2):435–439

Schwarz M (1973) The use of waste water in gravel culture. Proc 3rd Int Congr Soilless Culture, PO Box 52, Wageningen, pp 219–225

Schwarz M (1984) Daily course of root respiration under normal and saline conditions. Proc 6th Int Congr Soilless Culture, PO Box 52, Wageningen, pp 559–564

Schwarz M (1985) The use of saline water in hydroponics. Soilless Cult 1(1):25–34

Schwarz M, Gale J (1981) Maintenance respiration and carbon balance of plants at low levels of sodium chloride salinity. J Exp Bot 32:933–941

Schwarz M, Gale J (1983a) Growth response to salinity of high levels of carbon dioxide. J Exp Bot 35:193–196

Schwarz M, Gale J (1983b) The effect of heat and salinity stress on the carbon balance of *Xanthium strumarium*. In: Marcelle E (ed) Effects of stress in photosynthesis. Nijhoff and Junk, Dordrecht, pp 325–331

Schwarz M, Vaadia Y (1968) Limestone gravel as growth medium in hydroponics. Plant Soil 31:1

Sheldrake E Jr, Boodley JW (1965) Commercial production of vegetable and flower plants. Ext Bull 1065 Cornell Univ, Ithaca, NY

Shuval H (1969) Health factors in the reuse of waste water for agricultural purposes. Public Health Paper, World Health Organization, Jerusalem

Smith DL (1987) Rockwool in horticulture. Grower Books, London

Smith IE (1982) Soil cultivation of cucumbers and tomatoes under protection in Natal, PhD Thesis. Univ of Natal, Pietermaritzburg, SA

Smith PF (1962) Mineral analysis of plant tissues. Annu Rev Plant Physiol 13:81–108

Smolders E, Merckx R, Schooraerts F, Vlassak K (1991) Continuous shoot growth monitoring in hydroponics. Physiol Plant 83:83–92

Soffer H, Levinger D (1980) The Ein-Gedi system – research and development of a hydroponic system. ISOSC Proc 5th Int Congr Soilless Culture, PO Box 52, Wageningen, pp 241–252

Sonneveld C (1980) Growing cucumbers and tomatoes in rockwool. ISOSC Proc 5th Int Congr Soilless Culture, PO Box 52, Wageningen, pp 253–262

Sonneveld C (1981) A method for calculating the composition of nutrient solutions for soilless cultures. Proefstation voor Tuinbouw onder Glas, Naakldwijk, Informatiereeks No 57 (translated edition) 13 pp

Steiner AA (1968) Soilless culture – O_2 requirements for root growth of cuttings in water. Ann J Bot 17:842–61

Steiner AA (1976) Nomenclature with hydroponics. ISOSC 4th Int Congr Soilless Culture, PO Box 52, Wageningen, pp 19–20

Steiner AA (1980) The selective capacity of plants for ions and its importance for the composition and treatment of the nutrient solution. Acta Hortic 98:87–97

Steiner AA (1984) The universal nutrient solution. ISOSC Proc 6th Int Congr Soilless Culture, PO Box 52, Wageningen, pp 633–649

Stoughton RH (1969) Soilless cultivation and its application to commercial horticultural crop production. Food and Agriculture Organization of the United Nations, Rome, 61 pp

Stout PR (1961) Micronutrient needs for plant growth. Proc 9th Ann Calif Fertilizer Conf, pp 21–23

Suzuki Y, Shinohara Y, Shibuya M, Ikeda H (1984) Recent development of hydroponics in Japan. ISOSC Proc 6th Int Congr Soilless Culture, PO Box 52, Wageningen, pp 661–672

Terabayashi S, Takii K, Namiki T (1991) Variation in diurnal uptake of water and nutrients by tomato plants grown hydroponically. J Jpn Soc Hortic Sci 60(3):547–553

Tremblay N, Trudel MJ, Gosselin A (1984) Influence of supplementary lighting (HPS) on yield and mineral nutrition of tomato plants grown in hydroponic culture, ISOSC Proc 6th Int Congr Soilless Culture, PO Box 52, Wageningen, pp 697–703

Turban R (1978) CO_2 enrichment in hydroculture. Dtsch Gartenbau 32(16):666–667 (in German)

Van Goor BJ, de Jagar A, Voogt W (1988) Nutrient uptake by some horticultural crops during the growing period. ISOSC Proc 7th Int Congr Soilless Culture, PO Box 52, Wageningen, pp 163–176

Van Os EA (1982) Dutch developments in soilless culture. Outlook Agric 11:165–171

Van Os EA, Kuiken JCJ (1984) Mechanisation of lettuce growing in nutrient film technique. ISOSC Proc 6th Int Congr Soilless Culture, PO Box 52, Wageningen, pp 483–492

Verwer FL (1976) Growing horticultural crops in rockwool and nutrient film. ISOSC Proc 4th Int Congr Soilless Culture, PO Box 52, Wageningen, pp 107–119

Verwer FL, Welleman JJC (1980) The possibilities of Grodan rockwool in horticulture. ISOSC Proc 5th Int Congr Soilless Culture, PO Box 52, Wageningen, pp 263–277

Ward GM (1973) Leaf analysis for greenhouse vegetable crops. Ontario Ministry of Agriculture and Food AGDEX Factsheet 290, Ontario

Welleman JCC, Verber FL (1984) The technical installation for Grodan rockwool. ISOSC Proc 6th Int Congr Soilless Culture, PO Box 52, Wageningen, pp 763–772

Wilcox GE (1982) The future of hydroponics as a research and plant production method. J Plant Nutr 5(8):1031–1038

Wilson GCS, Hitchen GM (1984) The developments in hydroponic systems for the production of glasshouse tomatoes. ISOSC Proc 6th Int Congr Soilless Culture, PO Box 52, Waseningen, pp 793–800

Wilson GCS, Hall DA, McGregor AJ (1984) Perlite culture of tomatoes. West of Scotland Agricultural College, Auchincruwe, Tech Note No 219, 6 pp

Winsor GW (1982) Some nutritional aspects of nutrient film culture ("NFT"). Proc 9th Int Plant Nutrition Colloq, 2. Commonwealth Agricultural Bureaux, Farnham Royal, pp 722–727

Winsor GW, Adams P (1987) Diagnosis of mineral disorders of plants, vol 3. Glasshouse crops. Her Majesty's Stationary Office, London, 168 pp

Woolhouse HW (1981) Crop physiology in relation to agricultural production. In Johnson CB (ed) Physiological processes limiting plant productivity. Butterworth London, pp 1–33

Appendix

Appendix Table 1. Conversion factors (abbreviations in parentheses)

Multiply:	By factor:	To obtain:
Units of Length		
Centimeters (cm)	10	Millimeters (mm)
"	0.01	Meters (m)
"	0.3937	Inches (in)
"	0.3281	Feet (ft)
Feet	12	Inches
"	0.3333	Yards (yd)
"	30.48	Centimeters
"	0.3048	Meters
Inches	2.54	Centimeters
"	0.08333	Feet
"	0.02778	Yards
Meters	100	Centimeters
"	0.001	Kilometers (km)
"	39.37	Inches
"	3.281	Feet
"	1.094	Yards
Yards	3	Feet
"	36	Inches
"	0.9144	Meters
Miles	5280	Feet
"	1760	Yards
"	1.6093	Kilometers
Units of weight		
Kilograms (kg)	2.205	Pounds (lb)
"	1000	Grams (g)
Grams	1000	Milligrams (mg)
"	0.03527	Ounces (oz)
"	0.00220462	Pounds
Grams/1000 liters	0.0084	Pounds/1000 gallons
Milligrams	0.001	Grams
Pounds	0.454	Kilograms
"	16	Ounces
"	0.12	Gallons (gal)
Pounds/1000 gallons	133	Grams/1000 liters
Ounces	0.0625	Pounds
"	28.3495	Grams
Tons (t) (metric)	2205	Pounds

Appendix Table 1. *Continued*

Multiply:	By factor:	To obtain:
Tons (metric)	1000	Kilograms
Pounds of water	0.016	Cubic feet of water
"	27.68	Cubic inches of water
"	0.1198	Gallons of water
Units of area		
Acre (a)	43.560	Square feet (sq. ft.)
"	4840	Square yards (sq. yd.)
"	0.40468	Hectares (ha)
"	4046.8	Square meters (sq. m)
Hectares	2.471	Acres (a)
"	10.000	Square meters
Dunam	1000	Square meters
Square feet	144	Square inches (sq. in)
" "	0.929	Square meters
Square mile (sq. mile)	640	Acres
" "	258.99	Hectares
Square yards	9	Square feet
" "	0.83613	Square meters
Units of Volume		
Liters (l)	0.03531	Cubic feet (cu. ft)
"	61.02	Cubic Inches (cu. in)
"	0.2642	Gallons (gal) (US)
"	1000	Milliliters (ml)
"	1000	Cubic centimeters (cc)
"	1.06	Quarts (US)
Gallons (US)	0.1337	Cubic feet
" "	3.785	Liters
" "	0.83267	Imperial gallons
" "	231	Cubic inches
Cubic centimeters	6.102×10^{-3}	Cubic inches
" "	1×10^{-6}	Cubic meters
" "	3.53×10^{-5}	Cubic feet
" "	2.654×10^{-4}	Gallons (US)
" "	1.057×10^{-3}	Quarts (US)
" "	1000	Liters
Milliliters	1000	Liters
Cubic feet	2.832×10^{4}	Cubic centimeters
" "	1728	Cubic inches
" "	7.48052	Gallons (US)
" "	28.32	Liters
Cubic inches	1.639×10^{-2}	Liters
" "	16.39	Cubic centimeters
" "	4.329×10^{-3}	Gallons (US)
" "	5.787×10^{-4}	Cubic feet
Cubic meters	264.2	Gallons (US)
" "	35.31	Cubic feet
Units of concentration		
Parts per million (ppm)	1	Milligrams per liter

Appendix Table 1. *Continued*

Multiply:	By factor:	To obtain:
Parts per million (ppm)	0.001	Grams per liter
" " "	0.1	Pounds per 12,000 gallons
" " "	1	Ounces per 7,400 gallons
Milligrams per liter (mg/l)	1	Parts per million
Grams per liter (g/l)	1000	Parts per million
$EC^a \times 10^3$	10	Milliequivalents per liter (mEq/l)
" " "	640	Parts per million
" " "	0.36	Osmotic pressure (OP) (in the range 3–30 mmhos/l)

[a] $EC \times 10^3$ = electrical conductivity in millimhos (mmhos) per centimeter.

Units of temperature

$*F = (*C \times \frac{9}{5}) + 32$

$*C = (*F - 32) \times \frac{5}{9}$

Units of illumination

Units	Lx	Phot	Milliphot	Foot-candle
1 lux	1	0.0001	0.1	0.09
1 phot	10 000	1	1000	929
1 milliphot	10	0.001	1	0.929
1 foot-candle	10.76	0.001	1.07	1
1 lumen cm^{-2}	10^4	1	1000	929

Conversions, energy per unit area

	Joule cm^{-2}	BTU ft^{-2}	watt-h m^{-2}	g-cal cm^{-2}
1 Joule cm^{-2}	1	0.881	2.78	0.239
1 BTU ft^{-2}	1.136	1	3.15	0.271
1 watt-h m^{-2}	0.3597	0.317	1	0.086
1 g-cal cm^{-2}	4.19	3.69	11.624	1

Units of power per unit area (intensity)

	erg s^{-1} cm^{-2}	Langley min^{-1}	g-cal min^{-1} cm^{-2}	BTU h^{-1} ft^{-2}	watt cm^{-2}
1 erg s^{-1} cm^{-2}	1	1.43×10^{-6}	1.43×10^{-6}	6.47×10^{-9}	10
1 Langley min^{-1}	6.99×10^5	1	1	221.13	0.0698
1 g-cal min^{-1} cm^{-2}	6.99×10^5	1	1	221.13	0.0698
1 BTU h^{-1} ft^{-2}	1.54×10^8	4.52×10^{-3}	4.52×10^{-3}	1	3.16×10^{-4}
1 watt cm^{-2}	0.1	14.32	14.32	3.16×10^3	1
1 watt m^{-2}	1000	1.43×10^{-3}	1.43×10^{-3}	3.17×10^2	10^{-4}

Appendix Table 2. Amounts of chemicals to be added to supply 1 mmol of nutrient or nutrient radical

Chemicals	NO_3	P	SO_4	Cl	K	Mg	Ca	NH_4
	g/1000 l							
KNO_3	110	–	–	–	110	–	–	–
$Ca(NO_3)_2$	90	–	–	–	–	–	180	–
$NaNO_3$	90	–	–	–	–	–	–	–
$(NH_4)_2SO_4$	–	–	140	–	–	–	–	70
NH_4NO_3	80		–	–	–	–	–	80
$NH_4H_2PO_4$ (Ammophos)	–	140	–	–	–	–	–	140
$NH_4H_2PO_4$ (technical)	–	120	–	–	–	–	–	120
$(NH_4)_2HPO_4$ (technical)	–	140	–	–	–	–	–	70
KH_2PO_4	–	140	–	–	140	–	–	–
KCl	–	–	–	80	80	–	–	–
K_2SO_4	–	–	200	–	100	–	–	–
$CaH_4(PO_4)_2$ (technical)	–	135	–	–	–	–	270	–
$CaH_4(PO_4)_2$ (triple super)	–	155	–	–	–	–	310	–
$CaH_4(PO_4)_2$ (single super)	–	375	–	–	–	–	750	–
$MgSO_4 \cdot 7H_2O$ (Epsom Salts)	–	–	260	–	–	260	–	–
$MgSO_4$ (anhydrous)	–	–	130	–	–	130	–	–
$CaSO_4$	–	–	190	–	–	–	190	–
$CaCl_2$	–	–	–	75	–	–	150	–

Example: to supply 1 mmol of nitrate, add 110 g KNO_3 or 90 g calcium nitrate to 1000 l of water.

Appendix Table 3. Suggested weights (in g) of fertilizers for making up 1000 l of nutrient solution

	Nonsaline water				
Chemicals containing macroelements	Solution I	Solution II	Solution III	Solution IV	Saline water
Calcium nitrate	–	–	–	800	–
Sodium nitrate	750	–	–	–	–
Potassium nitrate	–	800	400	–	900
Potassium sulfate	750	–	400	–	–
Potassium phosphate	–	–	–	300	–
Magnesium sulfate	450	450	450	450	–
Triple superphosphate	350	–	350	–	–
Phosphoric acid	–	120	–	–	140
Calcium sulfate (gypsum)	120	200	–	–	–
Calcium ammonium nitrate	–	–	500	–	–
Ammonium sulfate	120	120	–	120	–
Chemicals containing microelements					
Ferrous sulfate	5.0				8.0
Iron sequestrene	Amounts given in text				–
Boric acid	3.0				1.5
or Borax	10.0				4.0
Zinc sulfate	0.2				0.3
Copper sulfate	0.2				0.3
Manganese sulfate	1.0				1.0

Appendix Table 4. Composition of nutrient solutions used at commercial and experimental stations (approximate figures) (g/1000 l water). (Schwarz 1968)

Chemicals	Commercial solution (tap water) in Israel	Commercial solution (saline water) in Israel	N. Jersey Expt. Stat. (Turner)	Ohio State Expt. State WP Solution (Kiplinger)	Naaldwijk Expt. Stat. Holland (Steiner)	Weihenstephan Expt. Stat. Germany (Penningsfeld)	Weihenstephan Expt. Stat. Germany (Luck)	Cal. Expt. Stat. (Hoagland)	Cal. Expt. Stat. (commercial) (Ellis)
Calcium nitrate	–	–	2430	–	900	868	692	90	720
Sodium nitrate	750	–	–	–	–	–	–	–	–
Potassium nitrate	–	900	–	608	440	416	315	540	660
Potassium sulfate	750	–	–	–	–	–	–	–	–
Monopotassium phosphate	–	–	285	–	–	284	–	–	–
Monoammonium phosphate	–	–	–	–	–	–	–	–	120
Monocalcium phosphate	–	–	–	282	–	–	–	140	
Patentkali 28% K_2O	–	–	–	–	400	–	346		
Magnesium sulfate	450	[a]	570	511	150	378		130	520
Superphosphate	–	–	–	–	–	–	692	–	–
Triple superphosphate	–	–	–	–	250	–	–	–	–
Phosphoric acid	120	140	–	–	–	–	–	–	–
Calcium sulfate	200	[a]	–	1214	–	–	–	–	–
Ammonium sulfate	120	50	150	110	10	10	–	–	–
Ferrous sulfate	5	8	–	2.5–5	–	20	14	14	–
Boric acid	–	1.5	–	1.4–2.8	–	–	–	–	–
Borax	10	[a]	–	–	10	10	5	1.7	–
Zinc sulfate	0.2	0.3	–	0.1–0.2	0.04	0.04	0.1	0.8	–
Copper sulfate	0.2	0.3	–	0.1–0.2	0.04	0.04	0.1	0.6	–
Manganese sulfate	1.0	1.0	–	1 –2	5	5	2.5	2	–
Sodium molybdate	–	–	–	–	0.125	–	–	–	–
Total salt content	2406	1108	3435	2725	2129	1991	3066	919	2020

[a]Natural brackish water contains a sufficient quantity of this element.

Appendix Table 5. Properties of chemicals containing macroelements

Chemical	Molecular weight	Approx. content (in %) of nutrient element(s) (corrected for usual impurities)	Physiological acid or base	Ratio of solvent to water	Storage
Potassium nitrate (saltpeter) KNO_3	101.10	13-N, 36-K	Neutral	1:4	Dry
Calcium nitrate $Ca(NO_3)_2$	164.10	16.5-N, 23.5-Ca	Neutral	1:1	Closed and dry (very hygroscopic)
Sodium nitrate (Chile saltpeter) $NaNO_3$	85.01	15.5-N	Neutral	1:1	Dry
Ammonium sulfate $(NH_4)_2SO_4$	132.15	20.5-N	Acid	1:2	–
Ammonium nitrate $(NH_4)NO_3$	80.05	40-N	Acid	1:1	Dry and cold
Ammonium dihydrogen phosphate $NH_4H_2PO_4$	115.04	10-N, 22-P	Acid	1:4	–
Ammonium monohydrogen phosphate $(NH_4)_2HPO_4$	132.07	12-N, 22-P	Base	1:2	–
Monopotassium phosphate KH_2PO_4	136.09	28-K, 22.5-P	Acid	1:3	–
Dipotassium phosphate K_2HPO_4	174.18	–	Base	1:1	–
Potassium chloride KCl	74.55	49-K	Neutral	1:3	Closed and dry
Potassium sulfate K_2SO_4	174.26	40-K	Neutral	1:15	–
Monocalcium phosphate $Ca(H_2PO_4)2{\cdot}H_2O$	252.09	–	Acid	1:60	–
Single superphosphate $CaH_4(PO_4)_2$	Variable	16.20-P, 12-Ca	Acid	1:400	–
Triple superphosphate $CaH_4(PO_4)_2$	Variable	32-40-P, 5-Ca	Acid	1:300	–
Dicalcium phosphate $CaPHO_4{\cdot}2H_2O$	172.10	–	Base	1:5000	–
Magnesium sulfate (Epsom salts) $MgSO_4{\cdot}7H_2O$	246.50	9.5-Mg	Neutral	1:2	–
Magnesium sulfate (anhydrous) $MgSO_4$	120.39	10.5-Mg	Neutral	1:10	–
Magnesium chloride $MgCl_2{\cdot}6H_2O$	203.33	–	Neutral	–	Closed and dry (very hygroscopic)
Calcium sulfate (gypsum) $CaSO_4{\cdot}2H_2O$	172.18	22-Ca	Neutral	1:500	
Calcium chloride $CaCl_2{\cdot}6H_2O$	219.09	20-Ca	Neutral	1:1	Closed and dry (very hygroscopic)

Appendix Table 6. Properties of chemicals containing microelements

Chemical	Formula	Molecular wt.	Physiological acid or base	Solubility g/100 ml water	Storage
Ferrous sulfate	$FeSO_4 \cdot 7H_2O$	278.03	acid	21	–
Ferric chloride	$FeCl_3 \cdot 6H_2O$	270.32	acid	45	Brown bottle
Ferrous ammonium sulfate	$FeSO_4(NH_4)_2SO_4 \cdot 6H_2O$	392.16	acid	21	Closed and dry
Boric acid	H_3BO_3	61.84	acid	4.9	–
Sodium tetraborate (Borax)	$Na_2B_4O_7 \cdot 10H_2O$	381.43	base	3.7	–
Copper sulfate	$CuSO_4 \cdot 5H_2O$	249.69	acid	18.7	–
Copper chloride	$CuCl \cdot 2H_2O$	170.49	acid	42.2	Closed and dry
Manganese sulfate	$MnSO_4 \cdot 4H_2O$	223.06	acid	38	–
Manganese chloride	$MnCl_2 \cdot 4H_2O$	197.91	acid	43	Dry (hygroscopic)
Zinc sulfate	$ZnSO_4 \cdot 7H_2O$	287.56	acid	36.6	–
Zinc chloride	$ZnCl_2$	136.29	acid	78	Closed and dry

Appendix Table 7. Conversion factors for fertilizer salts[a]

A	B	Conversion factor A to B	Conversion factor B to A
Nitrogen (N)	Ammonia (NH_3)	1.216	0.822
" "	Nitrate (NO_3)	4.426	0.226
" "	Ammonium sulfate ($(NH_4)_2SO_4$)	4.716	0.212
" "	Sodium nitrate ($NaNO_3$)	6.017	0.165
" "	Potassium nitrate (KNO_3)	7.214	0.139
Sodium nitrate ($NaNO_3$)	Sodium (Na)	0.270	3.690
" "	Ammonia (NH_3)	0.204	4.992
Phosphorus (P)	Phosphate (PO_4)	3.066	0.326
Phosphoric acid (P_2O_5)	Phosphorus (P)	0.437	2.289
" "	Monocalcium phosphate ($CaH_4(PO_4)_2 \cdot H_2O$)	1.648	0.607
Potassium (K)	Potash (K_2O)	1.205	0.830
"	Potassium nitrate (KNO_3)	2.559	0.386
"	Potassium sulfate (K_2SO_4)	2.228	0.449
"	Potassium chloride (KCl)	1.907	0.524
-	Potassium carbonate (K_2CO_3)	1.767	0.566
-	Potassium bicarbonate ($KHCO_3$)	2.580	0.388
Potash (K_2O)	Potassium nitrate (KNO_3)	2.147	0.466
"	Potassium sulfate (K_2SO_4)	1.851	0.540
"	Potassium chloride (KCl)	1.583	0.632
Calcium oxide (CaO)	Calcium carbonate ($CaCO_3$)	1.785	0.560
" "	Calcium sulfate ($CaSO_4$)	2.428	0.412
" "	Calcium (Ca)	0.715	1.399
" "	Tricalcium phosphate ($Ca_3(PO_4)_2$)	1.844	0.542
" "	Gypsum ($CaSO_4 \cdot 2H_2O$)	3.071	0.326
Calcium hydroxide ($Ca(OH)_2$)	Calcium (Ca)	0.541	1.849
" "	Calcium oxide (CaO)	0.757	1.321
Calcium carbonate ($CaCO_3$)	Calcium (Ca)	0.400	2.497
" "	Calcium hydroxide ($Ca(OH)_2$)	0.741	1.351
Magnesium sulfate ($MgSO_4$)	Magnesium oxide (MgO)	0.335	2.986
" "	Magnesium (Mg)	0.200	4.950
Magnesium oxide (MgO)	Magnesium (Mg)	0.603	1.666
Magnesium sulfate, technical $MgSO_4 \cdot 7H_2O$	Magnesia (MgO)	0.164	6.110
Sulfur (S)	Sulfuric acid (H_2SO_4)	3.059	0.327
Iron (Fe)	Iron sulfate ($FeSO_4 \cdot 7H_2O$)	4.980	0.200
Boron (B)	Borax ($Na_2B_4O_7 \cdot 10H_2O$)	8.807	0.114
Copper (Cu)	Copper sulfate ($CuSO_4 \cdot 5H_2O$)	3.930	0.255
Manganese (Mn)	Manganese sulfate ($MnSO_4 \cdot 4H_2O$)	4.060	0.250
"	Manganese oxide (MnO)	1.290	0.775

[a]In order to find the equivalents of items in column "A" in terms of those in column "B", multiply the amount of "A" by factor in column "A to B". For reverse procedure multiply amount of "B" by factor "B to A".

Appendix Table 8. Ranges of element and ions present in nutrient solutions, water and plants (in ppm except where otherwise indicated). (Schwarz 1968)

	Nutrient solution prepared from regular tap water			Nutrient solution prepared from saline water			Amount usually present in tap water	Amount in saline water		Amount in plants (dry weight)	
	Insufficient amount	Optimum	Maximum or excess	Insufficient amount	Optimum	Maximum or excess		Medium ranges	High ranges	Usually present	High ranges
Nitrate (NO_3^-)	200	300–900	1000	300	300–900	1000	0–8			0.2 –4.0%	
Ammonium (NH_4^+)	–	0–40	100	–	0–40	100					
Phosphorus (P)	30	30–90	100	30	50–100	150	0–5			0.3 –3.0%	
Potassium (K^+)	150	200–400	600	150	200–400	600	0–15			0.2 –3.5%	
Calcium (Ca^{2+})	100	150–400	600	100	150–400	600	5–50	50–100	200	0.1 –10.0%	
Magnesium (Mg^{2+})	25	25–75	50	25	50–100	200	0–10	10–60	140	0.05–1.0%	
Sulfate (SO_4^{2-})	150	200–1000	1000	200	200–1000	1200	0–20	20–200	900	0.2 –2.0%	
Chloride (Cl^-)	30	–350	500	30	700–1000	1500	5–300	300–600	1500	0.2 –1.0%	
Sodium (Na^+)	–	–	400	–	400–500	800	1–200	200–400	800	0.1 –1.0%	to 15%
Iron (Fe^{2+})		0.5–2			0.8–3		0.05–1.0			20 –100	to 1500
Boric acid (H_2BO_3)		0.2–1	5		0.2–1	5	0.1 –0.5	1–10	20	10 –100	to 600
Zinc (Zn^{2+})		0.2–2	20		0.8–4	20	1			5 –75	to 600
Copper (Cu^{2+})		0.1–2	5		0.8–3	10	0.05–0.5			1 –25	
Manganese (Mn^{2+})		1 –5	15		1 –5	15	10			5 –50	to 800
Cobalt (Co^{2+})							0.01			0.01–1.0	
Fluoride (F^-)							0.05–1.0	5–10	15	–1.0	to 800
Molybdenum (Mo^{2+})							0.01			0.01–20.0	

Note: dash indicates chemical is not present; blank indicates absence of information.

Appendix Table 9. Recommended pH for soilless-culture crops

Crop	pH range					
	5.0	5.5	6.0	6.5	7.0	7.5
Artichoke				—	—	
Bean		—	—	—		
Cabbage			—	—	—	
Celery			—	—	—	
Cucumber		—	—	—		
Eggplant		—	—			
Lettuce		—	—	—		
Onion			—	—		
Potato		—	—			
Pea			—	—	—	
Pepper		—	—	—		
Radish		—	—	—		
Rhubarb			—	—		
Spinach			—	—	—	
Tomato		—	—	—		
Watermelon	—	— - -	- -			
Asparagus			—	—	—	
Amaryllis	—	—	—			
Asparagus sprengeri		—	—	—		
Begonia	—	—	—			
Carnation			—	—		
Chrysanthemum	—	—	—	—		
Cyclamen			—	—	—	
Dahlia			—	—	—	
Dianthus		—	—	—		
Gerbera jamesonii		—	—	—		
Gladiolus		—	—	—		
Grape		—	—	—		
Sweet pea			—	—	—	
Rose		—	—	—		
Strawberry		—	—	—		
Violet			—	—	—	
Zinnia			—	—		

Appendix Table 10. Nutrient solution concentration recommended[a] for flowes, in NFT or rockwool culture

Macroelements in mmol l^{-1}	Carnation[b]	Roses[b]	Gerbera[c]	Anthurium andreanum[c]	Chrysanthemus[b]	Cymbidium[c]	Begonia[c]	Green plants[b]	Poinsettien[c,d]	Cyclames[a,d]
NO_3^-	11.5	10.5	9.0	7.5	10.5	8.0	9.0	10	11.5	11.0
$H_2PO_4^-$	1.5	1.5	1.0	1.0	1.0	1.25	1.0	1.5	1.5	1.0
SO_4^{2-}	1	0.75	1.0	1.0	1.0	1.0	1.0	1.0	1.5	1.5
NH_4^+	0.25	0.25	0.75	0.5	0.5	1.25	0.5	0.5	0.25	0.25
K^+	6.25	5.75	5.75	6.0	5.0	4.0	6.0	5.5	7.25	5.75
Ca^{2+}	3.5	3.0	2.0	1.5	3.0	1.5	2.0	3.0	3.5	4.0
Mg^{2+}	0.75	0.75	0.75	0.5	1.0	1.0	0.75	0.75	0.75	0.5
Microelements in μ mol l^{-1}										
Fe	35	35	55	55.0	60	10	55	35	15	20
Mn	7.5	20	10	10.0	20	20	15	10	10	10
Zn	3	2.5	3	3.0	3	4	4	3	3	3.0
B	20	20	15	20.0	20	20	20	20	20	25
Cu	0.5	0.5	1	0.5	0.5	0.5	1	0.5	0.5	0.5
Mo	0.5	0.5	0.5	0.5	0.5	0.5	0.5	0.5	0.5	0.5

[a]C. Sonneveld, Naaldwijt Experimental Station, The Netherlands.
[b]Highly recommended.
[c]Recommended.
[d]Only in rockwool.

Appendix Table 11. Oxygen solubility (DO) in water exposed to water-saturated air[a]

	Chloride concentration in water (mg/l)					Difference per 100 mg chloride
	0	5000	10000	15000	20000	
Temp. (°C)	Dissolved oxygen (mg/l)					
0	14.6	13.8	13.0	12.1	11.3	0.017
1	14.2	13.4	12.6	11.8	11.0	0.016
2	13.8	13.1	12.3	11.5	10.8	0.015
3	13.5	12.7	12.0	11.2	10.5	0.015
4	13.1	12.4	11.7	11.0	10.3	0.014
5	12.8	12.1	11.4	10.7	10.0	0.014
6	12.5	11.8	11.1	10.5	9.8	0.014
7	12.2	11.5	10.9	10.2	9.6	0.013
8	11.9	11.2	10.6	10.0	9.4	0.013
9	11.6	11.0	10.4	9.8	9.2	0.012
10	11.3	10.7	10.1	9.6	9.0	0.012
11	11.1	10.5	9.9	9.4	8.8	0.011
12	10.8	10.3	9.7	9.2	8.6	0.011
13	10.6	10.1	9.5	9.0	8.5	0.011
14	10.4	9.9	9.3	8.8	8.3	0.010
15	10.2	9.7	9.1	8.6	8.1	0.010
16	10.0	9.5	9.0	8.5	8.0	0.010
17	9.7	9.3	8.8	8.3	7.8	0.010
18	9.5	9.1	8.6	8.2	7.7	0.009
19	9.4	8.9	8.5	8.0	7.6	0.009
20	9.2	8.7	8.3	7.9	7.4	0.009
21	9.0	8.6	8.1	7.7	7.3	0.009
22	8.8	8.4	8.0	7.6	7.1	0.008
23	8.7	8.3	7.9	7.4	7.0	0.008
24	8.5	8.1	7.7	7.3	6.9	0.008
25	8.4	8.0	7.6	7.2	6.7	0.008
26	8.2	7.8	7.4	7.0	6.6	0.008
27	8.1	7.7	7.3	6.9	6.5	0.008
28	7.9	7.5	7.1	6.8	6.4	0.008
29	7.8	7.4	7.0	6.6	6.3	0.008
30	7.6	7.3	6.9	6.5	6.1	0.008

[a]At a total pressure of 760 mm Hg.

Appendix Table 12. Relative salinity tolerance of crops arranged according to decreasing tolerance within groups[a]

Tolerant (12–8 mmho/cm)	Moderately tolerant (8–4 mmho/cm)		Sensitive (3–2 mmho/cm)
Garden beets	Tomato	Pepper	Field beans
Kale	Broccoli	Onion	Radish
Asparagus	Cabbage	Broadbean	Celery
Spinach	Cauliflower	Peas	Green beans
	Lettuce	Squash	Strawberry
	Potatoes	Cucumber	
	Grape		

[a]Figures taken from *Saline and Alkali Soils*, US Salinity Laboratory Staff, Agricultural Handbook No. 60, p. 160.

Appendix Table 13. Relative tolerance of crops to boron[a]

Tolerant (4 ppm)	Semitolerant (2 ppm)	Sensitive (1 ppm)
Asparagus	Sweet pea	Grape
Broadbeen	Potato	
Onion	Tomato	
Cabbage	Radish	
Lettuce	Pumpkin	
Gladiolus	Bell pepper	
	Zinnia	
2 ppm[b]	1 ppm	0.3 ppm

[a]From *Saline and Alkali Soils*, US Salinity Laboratory Staff, Agricultural Handbook, No. 60, p. 163.
[b]Indicates limits of tolerance for boron in irrigation waters.

Appendix Table 14. Approximate molecular proportion of elements required by flowers (according to F. Penningsfeld). (Schwarz 1968)

	N	P_2O_5	K_2O	CaO	MgO	Fe	Mn	B	Cu	Zn	Percentage of total salt concentration of nutrient solution: Optimum	Maximum (small depression in yield)
Adiantum	1	0.5	1.5								0.05–0.1	0.25
Aechmea fasciata	1	1.6	3								0.3 –0.3	0.5
Anemone coronaria											0.2 –0.3	0.6
Anthurium andreanum						× 2					0.25	
Anthurium scherzerianum (vegetative)	6	1	2				× 2				0.1	
Anthurium scherzerianum (flowering)	3	1	4								0.1	0.5
Aphelandra squarrosa	2	0.8	1.5								0.25	
Asparagus plumosus	3	1	2								0.1 –0.2	0.6
Asparagus sprengeri	3	1	5								0.2 –0.7	0.4
Azalea indica	3	2	2			× 2					0.1 –0.2	
Begonia	2	1	1								0.2	
Camellia japonica	2	0.8	1.5			× 2					0.1 –0.2	
Chrysanthemum indicum	1	0.8	1.4							× 2	0.2 –0.6	0.8
Cattleya mossiae	3	0.8	1.5								0.2 –0.3	0.7
Cissus antarctica	2	0.8	1.5								0.3 –0.5	
Cyciamen	2	0.8	2.2	0.7	0.4	× 2		× 2		× 2	0.3 –0.4	0.6
Dendrobium phalaenopsis	1	0.8	2.0							× 2	0.2	
Dianthus (winter)	1	0.8	2.3			× 2		× 2		× 2	0.4 –0.8	0.8
Dianthus (summer)	1	0.8	1.5								0.4 –0.8	0.8
Erica gracilis	3–4	1	2			× 2					0.2 (young 0.1)	
Euphorbia pulcherima	1	1.6	1.5								0.5	
Freesia	1	0.8	2.1			× 2		× 2	× 2	× 2	0.1 –0.3	0.6
Gardenia grandiflora	2	0.8	1.5	0.5	0.4	× 2					0.1	

Gerbera jamesonii	2	0.8	1.5			× 2		× 2	× 2	0.2	0.5
Gladiolus										0.3	
Lathyrus odoratus	1	0.5	2.2	0.6	0.4					0.2	0.4
Monstera deliciosa	2.5	0.8	1.5	0.7	0.1					0.3 –0.5	
Pelargonium (young)	1	0.9	1							0.5	
Pelargonium (older, flowering)	1	2	5							0.5	
Phalaenopsis	2	1	1	0.6	0.5	× 2				0.1	0.1
Primula obconica	1	0.8	1.5	2.1	0.3	× 2		× 2		0.1 –0.2	0.3
Ranunculus										0.2 –0.7	
Rosa	1	0.8	1.5	1.3	0.3					0.2	
Saintpaulia ionantha	2	0.8	1.6	1 0.3			× 2				
Sinningia speciosa	1	0.3	2	0.9	0.3					0.3 –0.6	
Vriesea splendens	1	0.5	2							0.3 –0.4	0.6
										0.05–0.1	0.3

Note 1: In this table, P is expressed as P_2O_5, K Ca, and Mg are also given as oxides. To convert P_2O_5 to P multiply it by 0.4; K_2O to K by 0.8; CaO to Ca by 0.7; MgO to Mg by 0.6.

Note 2: × 2 indicates that for this plant the quantity of the element to be added to the nutrient solution should be double that added to the standard solution.

Appendix Table 15. Optimal nutrient concentration in sand or sand/turf culture (according to F. Penningsfeld). (Schwarz 1968)

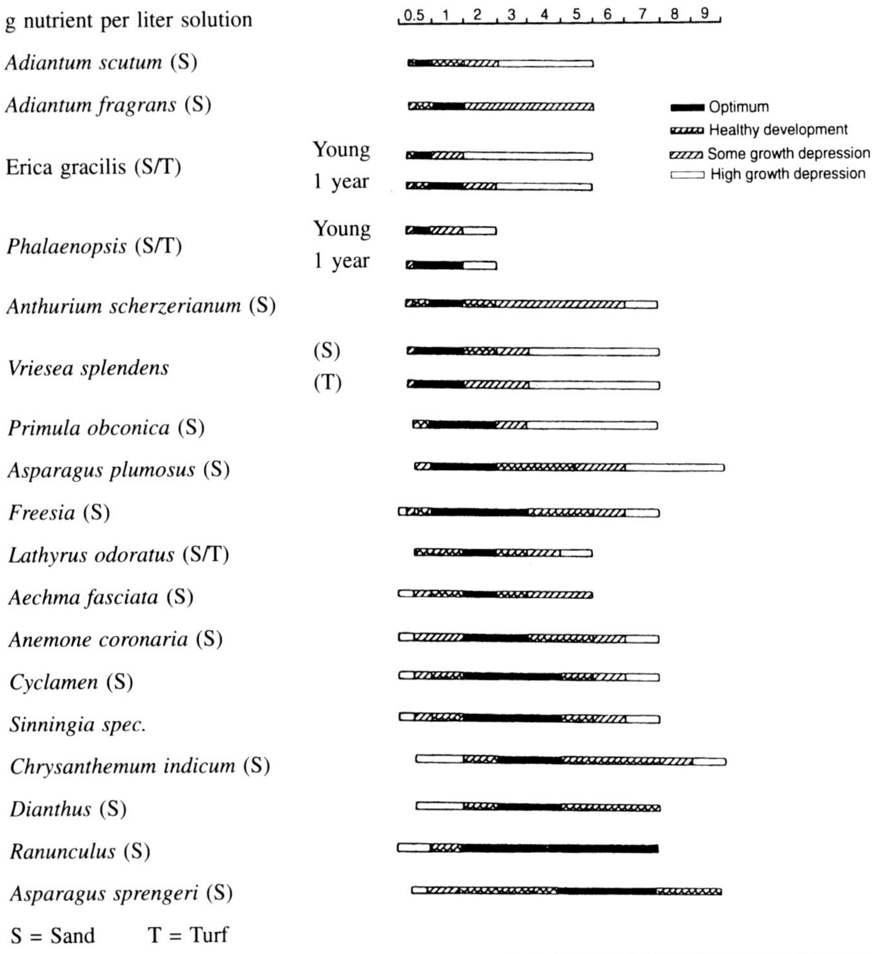

Appendix Table 16. Relationship between deficiencies and excesses of the various elements (according to Runken). (Schwarz 1968)

N	P	K	Ca	Mg	Fe	B	Mn	Cu	Zn	May simulate excess symptoms of
	d	d	e	e		d	d	d	d	Nitrogen
d		d	e	e	d			d	d	Phosphorus
d	d		d	d	e	d	d		d	Potassium
d	d	d		d	d	d	d	d	d	Calcium
	e	d	d							Magnesium
	d	d				d	d+	d	e	Iron
		d	d+	d						Boron
			d	d	d+					Manganese
					d		d		d	Copper
					d					Zinc
		d	d	d			d			Sodium
										May simulate deficiency symptoms of
	e	e		d						Nitrogen
			e	e						Phosphorus
	e		e	e	d		e			Potassium
		e		e			e			Calcium
	d		e		d					Magnesium
	e						e			Iron
e					e+					Manganese

Note: + Indicates the most expected and/or the most pronounced symptom.
Excess = e; Deficiency = d.

Appendix Table 17. Plant detriments

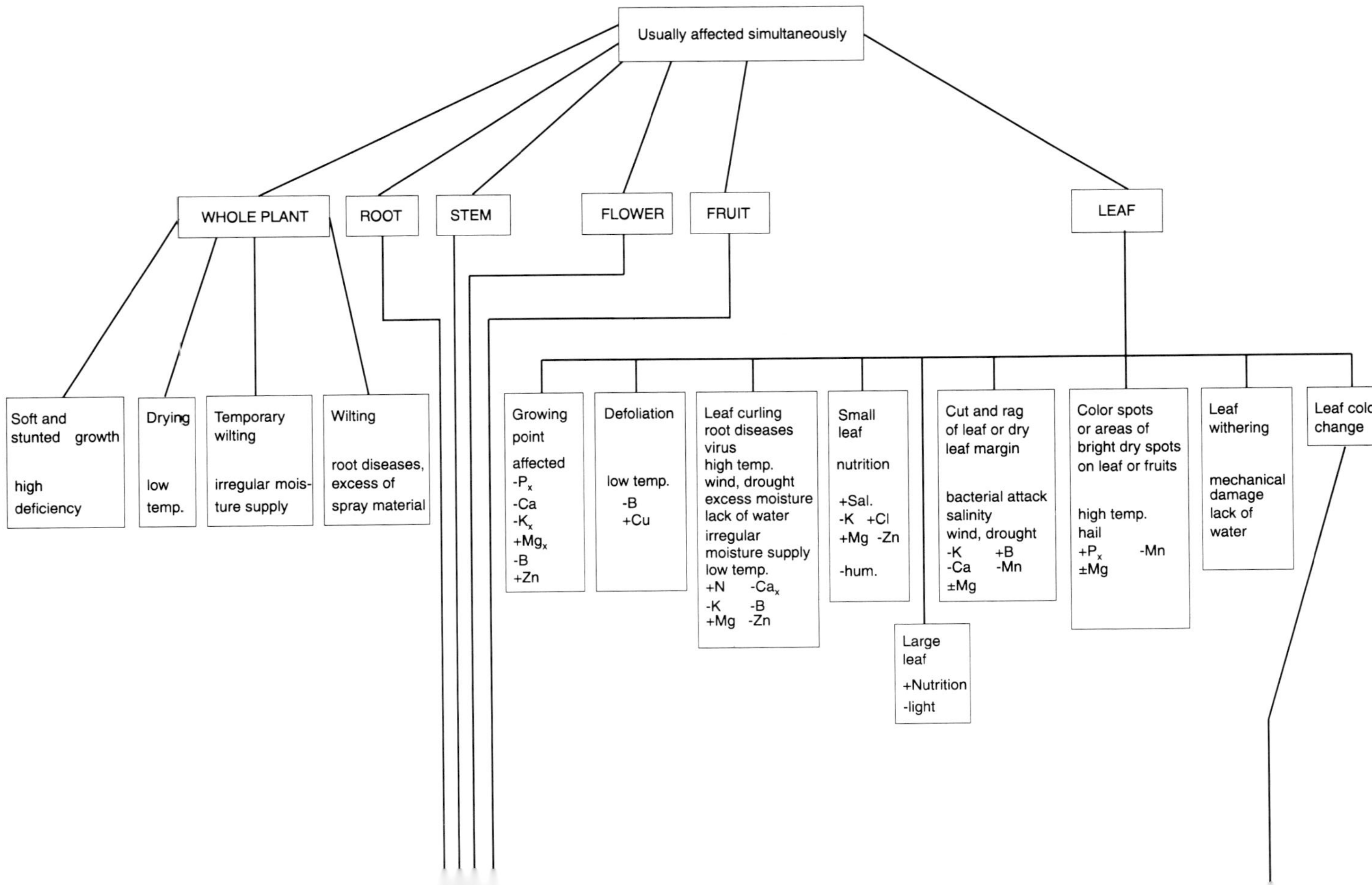

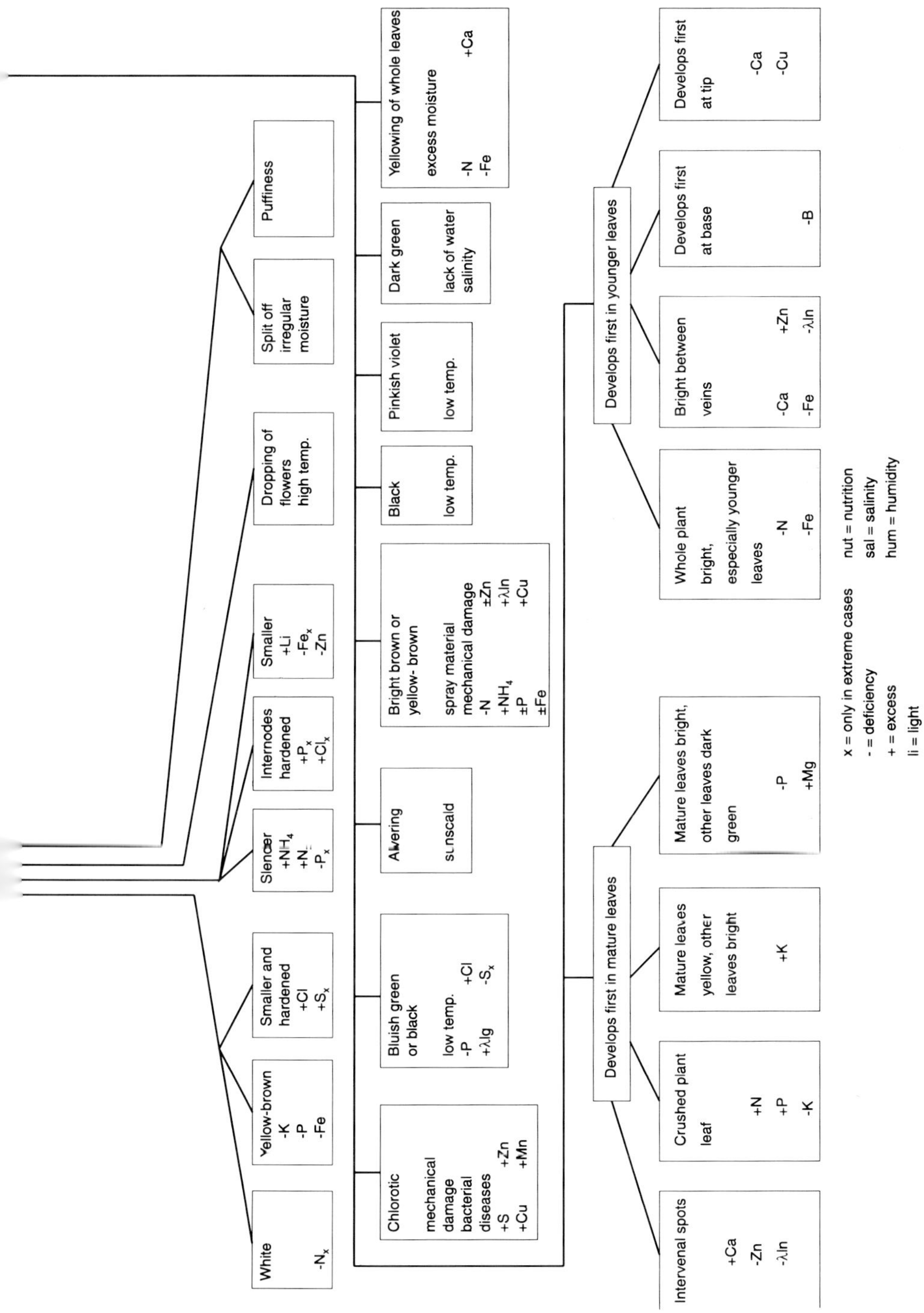
White
-N_x
Yellow-brown
-K
-P
-Fe
Smaller and hardened
+Cl
+S_x
Slender
+NH_4
+N
-P_x
Internodes hardened
+P_x
+Cl_x
Smaller
+Li
-Fe_x
-Zn
Dropping of flowers
high temp.
Split off
irregular moisture
Puffiness
Chlorotic
mechanical damage
bacterial diseases
+S
+Cu
+Zn
+Mn
Bluish green or black
low temp.
-P
+λlg
+Cl
-S_x
sunscald
Bright brown or yellow- brown
spray material
mechanical damage
-N
+NH_4
±P
±Fe
±Zn
+λln
+Cu
Black
low temp.
Pinkish violet
low temp.
Dark green
lack of water
salinity
Yellowing of whole leaves
excess moisture
-N
-Fe
+Ca
Develops first in mature leaves
Intervenal spots
+Ca
-Zn
-λln
Crushed plant leaf
+N
+P
-K
Mature leaves yellow, other leaves bright
+K
Mature leaves bright, other leaves dark green
-P
+Mg
Develops first in younger leaves
Whole plant bright, especially younger leaves
-N
-Fe
Bright between veins
-Ca
-Fe
+Zn
-λln
Develops first at base
-B
Develops first at tip
-Ca
-Cu
x = only in extreme cases
- = deficiency
+ = excess
li = light
nut = nutrition
sal = salinity
hum = humidity

Appendix Table 18. Physiological disorders of vegetable plants caused by mechanical or climatic factors

Main cause	Secondary cause	Part of plant affected	Kind of disorder	Species affected	Symptoms sometimes similar to those caused by	
					deficiency of	excess of
1. Sunscald	Leaves laid bare by wind; after trimming branches	Leaf; only underside affected	Silvering	Frequently: tomato, melon, water-melon Seldom: pepper	P	
2. Temporary high temperature	Low wind; fall of leaves due to leaf or root diseases which	Leaves	Rolled-up leaves during hot hours	Frequently: tomato		
	may be a result of dryness of media	Fruit	Bright dry spots; usually somewhat depressed	Seldom: melon, pepper, etc.		
3. Light deficiency	Too much shade; near building	Whole plant	Soft and stunted plant, large internodes	All plants		
4. Wind, drought		Leaf	1) Dry leaf margin; leaf curling 2) Cut and rag of plant leaf 3) General appearance of damaged plant	Frequently: tomato, onion, carrot. Seldom: squash, pumpkin, eggplant	P.K.	N
5. Hail		Leaf, stem, fruit	Bright and depressed spots on the tissue	Frequently: in soft plants such as tomato, pepper and cucumber		
6. Excess moisture	Inadequate root aeration	Leaf	Yellowing, first in younger leaves; leaf often curled	All plants		
7. Lack of water	Wind, saline water	Leaf	Dark green color, withering, leaf curling	All plants		

8. Irregular moisture supply	Often heavy rainfall, irregular irrigation		Temporary wilting; splitting of fruit	Tomato, etc.	
9. High temperature		Leaf, flower	Leaf curling, dropping of tomato flowers	Tomato, etc.	
10. Low temperature, above freezing point		Leaf, flower	Leaf curling; leaf color, pinkish violet, especially on veins (not estimated if yield depression occurs)	Tomato, pepper, carrot, cabbage	N,P,K
11. Low temperature, below freezing point		Leaf, whole plant	Drying of plant, withering black leaves and fruits	Cucumber, squash, melon, etc. tomato, pepper, eggplant, etc.	
12. Excessively high or low temperature		Fruit	Puffiness; fruit inadequately fleshed	Tomato	
13. Mechanical damage	Leaf-rubbing, wind	Leaf	Leaves withering, bright brown color	Cucumber, squash, melon, etc.	
14. Spray material excessive leaf fertilizing	Ammonium sulfate sulfur spray	Leaf	Leaf burned; bright brown color	Cucumber, squash, melon, etc.	

Subject Index

Acidity (see pH)
Aeration 5
– in NFT 73–75
– in root 118
– in water culture 5
Airoponics 4, 81–83
Aluminium 18
Ammonium see Nitrogen
Arid zones (see saline water)
Automatic Fertilizer 142

Bag culture 83–90
– growth media 84
Bark 39
Beds 6, 66
– in gravel culture 56–64
– in sand culture 31, 40–43, 64–66
– size 34, 61–62
– through line with polyethylene 65
Benches (see beds)
Bibliography 165
Bicarbonate 17
Biochemical Oxygen Demand – BOD 104
Blossom end rot 11, 14, 101, 153
(see also Diseases)
Boron 11, 15
– deficiency and toxicity 15
– in plants 183
– in saline water 101
(see also Microelements and Nutrient solution)
Brackish water (see Salts, Saline)
Bromine 17, 162

Calcium compounds 13
Calcium ion
– deficiency and toxicity 13–14
– in plant 14
– sources 13
(see also Nutrient solution)
Carbon dioxide (CO_2) 108–116
– controllers 145
– economics 116
– effects of high 112
– fruit development 111
– in salinity 102, 103
– leaf crops 110
– monitoring 114
– reproductive development 111
– root crops 110
– sources 113
– transpiration rate 94, 136
– vegetative growth 110
– yields in 115
Carnation 156
Chelating agents 14, 27
Chemical Oxygen Demand (COD) 104
Chloride ion 16
– deficiency and toxicity 16–17
(see also Microelements, Salts, Saline, Sewage, Nutrient solution)
Chrisanthemum 156
– in NFT 73
Climate (see Environment)
Closed systems 4, 56, 66, 67
Cobalt 17
Commercial 158
– considerations 159
– establishing units 158
– investment 162
– management 161
Conductivity
– EC Controlling 144
– determination 144–145
(see also Salts, Saline and Containers – Beds)
Controlling systems 131
– controllers 131, 145
– carbondioxide 145
– conductivity 144
– dissolved oxygen 145
– heat 137
– light 134, 136, 138–141
– sensors 137

Controlling systems (*Contd.*)
– temperature 135, 136
Conversion factors 171–173, 178
Copper 16
– deficiency and toxicity 16
– in salinity 102
(see also Microelements, Nutrient solution)
Cucumber 153
– CO_2 in 112
Culture methods 33–91
– airoponics 81–83
– bag culture 83–90
– gravel culture 56–65
– nutrient film culture 66–76
– polyurethane – as rock wool culture
– rockwool culture 45
– sand culture 40
– tray container technique 60
– vertical culture 90–91
– water culture 77–83

Deficiencies and Toxicities 18, 19, 28, 29, 187–191
(see under various ions 9–17)
– of Boron 16
– of calcium 13–14
– of chlorine 17
– of copper 17
– of iron 15

Deficiencies and Toxicities 18, 19, 28, 29, 187–191
(see under various ions 9–17)
– interaction 19
– of magnesium 13
– of manganese 15
– of nitrogen 10, 11
– of phosphorus 12
– of potassium 12
– of sulfur 13
– of zinc 16
Disease of plants 187, 188–191
– control 129
– deficiency and toxicity symptoms (see also various ions)
– in NFT 74
– in roots 129
Dissolved oxygen 145, 182
– controller 145
– in water 182
Drought 103

Economics (see Commercial)
EC control (see Conductivity)
EDTA see Chelating agents 27
Elements in plants 7, 179, 184
(see various ions)
Environment 101
(see also Carbon dioxide)

Ferric and Ferrous (see Iron)
Fertilizer 27, 88
– applications 143
Floating hydroponic systems 78
– Japanese systems 79
Flowers 156, 186
– elements required 184
– high CO_2 112
– optimal nutrient concentration 186
Foam plastic (see Media-plastic)
Foliar nutrition 30
Frost 137
Future aspects 162, 163

Gerbera 156
Gravel 34
– beef sice 61
– calcareous 35
– characteristics 34, 35
– gravity feed system 58
– layout 61
– scoria 35
– surface irrigation 3, 50
– tubirrigation method 3
Gravel–sand differences 44

Harvest index 123
Home unit 60
Hydroponics 3, 4
– advantages of 3, 4
– definition of 3, 4, 33
– disadvantages 4, 6
– history 1
– types of 3

Installation
– building material asphalt 65
– building material concrete 62
– building material earth 63
– building material plastic 64
– coating 65
– commercial 158
– terracing 59, 60
Iron ion
– compounds 14, 36
– deficiency and toxicity 14, 15

– sequestrene 14
(see also Microelements and Nutrient solution)
Irrigation
– frequency 55, 96
– gravity-feed 58
– in NFT system 70, 72
– subirrigation 56
– Tensiometer 96

Labor 5
Lettuce 155
– CO_2 in 112
– in NFT 71, 72
Light 138
– duration 139, 140
– in crop canopy 139
– in glass house 141
Liquid fertilizer
– apparatures 4

Macroelements 3
(see Calcium, Magnesium, Nitrogen, Phosphorus, Potassium, Sulfur)
Maintenance respiration 102, 103, 117, 121, 124
Manganese 15
– deficiency and toxicity 15, 16
Magnesium ion 13
– deficiency 13
– in plant 13
Management 161
Media 34
– airation 39
– bark 39, 85
– Kontan 37
– organic 37–40, 44
– peat 38–40, 88
– perlite 37, 87
– sawdust 39, 84
– size 35, 44
– sterilizing 130
– vermiculite 85
– wood bark 85
Media sand–gravel differences 44
Microelement 3, 14, 177
– in solution 22–24
(see Boron, Chlorine, Copper, Iron, Manganese, Molybdenum, Zinc)
(see also Nutrient solution lists)
Mobility of ions
(see at the various macroelements)
Molybdenum 17
(see also Microelements and Nutrient solution)

Nitrate ion 10
(see Nitrogen)
Nitrite ion 55
Nitrogen 9
– deficiency 10, 188
– functions of 9, 10
– in nutrient solution 10
– in plant 10, 11
– sources 10
– toxicity 11
– uptake 10, 19
(see also nutrient solution)
Non required elements 17
Nutrients 7
– absorbtion 119
– concentration and yield 29
– interactions 19, 127
– mobile and immobile 126, 127
(see nutrient solution and various ions)
Nutrient Film Techniques NFT 3, 66
– irrigation 67–69
– IMAG system 71
– nutrition 181
– plant supporters 69, 70
Nutrient solution 20–28
– analysis 31
– balance 4
– concentration 23–28, 29
– elements essential 7–9
– elements nonessential 17
– expressing 20
– foliar 30
– for flowers 181
– in gravel culture 56–61
– interaction 19
– lists of nutrient solutions 21, 22, 24, 52, 53, 54, 75, 77, 98, 179, 181, 186
– in NFT 75–77
– preparation of 20, 23–29
– in rock wool 51–52
– saline – see Salts 98, 179
– Steiner's formula 26
– testing of 31
– warming of 137
– water, tap 179

Open systems 4, 40, 45, 83
Osmosis 93, 97
Osmotic pressure 25, 93, 99, 100

Oxygen absorbtion 118–119
– in NFT 73
see Dissolved oxygen 145, 182
Orchids 157

Peat 38–40, 88
pH 28, 76, 180
– optimum for plants 180, 181
– in rock wool 51
Phosphorus ion 11
– deficiency and toxicity 12
– sources 11
– uptake 11
see also Nutrient solution
Plants
– adaptable to hydroponic 147
– analysis 127
– disorders 188–191
– salinity tolerance 183
– stresses 103
– transplanting 149
Photosynthesis
– efficiency 123
– photosynthetically active radiation (PAR) 139
Plastic
– media 3, 36, 37
see also Installation
Plastoponics 4
Potassium ion
– deficiency and toxicity 12
– in nutrient solution 36
– in plant 12
– sources 12
see also Nutrient solution

Respiration 135, 136
Rock wool 3
– airation 73
– catchment tank 67
– culture 4, 45
– description 45, 68
– diseases 5, 74
– IMAG system 71
– irrigation 55, 69
– layout 47–50
– nutrient solution 50–53, 75, 77, 181
– plant relationship 46
– propagation 51
– plant supporters 69
– plant systems 71
– water relationship 46
Rhutner system 91
Root-mist technique 82
Root 6
– death in NFT 74, 129–130
– diseases 129
– functioning 117
– hairs 6, 119, 149
– nutrient absorbtion 119
– nutrient uptake 120–122
– oxygen 118, 119
Root-shoot ratio 124, 125, 149, 152
Root crown 36, 124, 137

Salts, Saline 5, 96
– adjusting 97
– advantages 97, 102
– balancing nutrient solution 98, 101
– carbon balance 102
– climatic effect 101
– conductivity 93
– determination 144
– effects of 102, 103
– ion uptake 101
– maintenance respiration 102
– plant grown in 98, 99, 183
– tolerance 99, 183
– toxicity 101
– yields in 102
Sand culture 3, 34, 40, 41
Sand dunes 43
Sand–gravel differences 44
Sand sprinkling system 41
Sewage 40, 104–107
– composition 106
– determination 104
– health factors 107
– nutrient solution in 106
– time of sewage use 105
Seeds
– germinating 148
– watering 148
Seedling 148
– light for 151
Silicon 17
Sodium ion 17
Sterilizing 43, 130
Strawberry
– in NFT 72
Surface irrigation method 3
Sulfur ions 13
– deficiency and toxicity 13
– sources 13

Temperature 135–138
– and respiration 135
– control of 136

– day/night 136
– heat suppliers 137
– influences on plant 136
– optimum 136
Through culture
– with organic substrate 44
Tomato 150
– CO_2 in 112
– deficiency and toxicity symptoms 11
– disorders in 152
– temperature 151–152
Trace elements see Microelements
Transpiration 93–95, 136
– and CO_2 94
– rate 94
– water stress 95
Transplanting 149
Tube irrigation method 3
Tuff see Gravel-scoria

Vermiculite 36, 85
Vermiculaponics 4, 36
Vertical hydroponics 90

Water 92
– absorbtion 93
– consumption 5
– brackish see Salts
– liquid culture 3
– sewage 104–107
– stress 95, 96, 103–104
Wastewater see Salt, Saline, Sewage
Water culture 66, 77

Zinc 16
– deficiency and toxicity 16
– in salinity 102
(see also Microelements and Nutrient solution)

Springer-Verlag and the Environment

We at Springer-Verlag firmly believe that an international science publisher has a special obligation to the environment, and our corporate policies consistently reflect this conviction.

We also expect our business partners – paper mills, printers, packaging manufacturers, etc. – to commit themselves to using environmentally friendly materials and production processes.

The paper in this book is made from low- or no-chlorine pulp and is acid free, in conformance with international standards for paper permanency.

Printing: Mercedesdruck, Berlin
Binding: Buchbinderei Lüderitz & Bauer, Berlin